T0351282

ALFRED NORTH
WHITEHEAD
PHILOSOPHER OF TIME

ALFRED NORTH
WHITEHEAD
PHILOSOPHER OF TIME

Rémy Lestienne

CNRS Paris, France

World Scientific

NEW JERSEY · LONDON · SINGAPORE · BEIJING · SHANGHAI · HONG KONG · TAIPEI · CHENNAI · TOKYO

Published by

World Scientific Publishing Europe Ltd.

57 Shelton Street, Covent Garden, London WC2H 9HE

Head office: 5 Toh Tuck Link, Singapore 596224

USA office: 27 Warren Street, Suite 401-402, Hackensack, NJ 07601

Library of Congress Cataloging-in-Publication Data
Names: Lestienne, Rémy, author. | Lestienne, Rémy. Whitehead, philosophe du temps.
Title: Alfred North Whitehead, philosopher of time / Rémy Lestienne, CNRS Paris, France.
Other titles: Whitehead, philosophe du temps. English
Description: New Jersey : World Scientific, [2022] | Includes bibliographical references and index.
Identifiers: LCCN 2021060210 | ISBN 9781800611771 (hardcover) |
    ISBN 9781800611788 (ebook) | ISBN 9781800611795 (ebook other)
Subjects: LCSH: Whitehead, Alfred North, 1861-1947. | Process philosophy.
Classification: LCC B1674.W354 L4713 2022 | DDC 192--dc23/eng/20220211
LC record available at https://lccn.loc.gov/2021060210

British Library Cataloguing-in-Publication Data
A catalogue record for this book is available from the British Library.

Cover design by Lali Abril

Whitehead, philosophe du temps
Original published in French by CNRS Éditions
Copyright © CNRS Éditions, Paris, 2020

Copyright © 2022 by World Scientific Publishing Europe Ltd.

All rights reserved. This book, or parts thereof, may not be reproduced in any form or by any means, electronic or mechanical, including photocopying, recording or any information storage and retrieval system now known or to be invented, without written permission from the Publisher.

For photocopying of material in this volume, please pay a copying fee through the Copyright Clearance Center, Inc., 222 Rosewood Drive, Danvers, MA 01923, USA. In this case permission to photocopy is not required from the publisher.

For any available supplementary material, please visit
https://www.worldscientific.com/worldscibooks/10.1142/Q0347#t=suppl

Typeset by Diacritech Technologies Pvt. Ltd.
Chennai - 600106, India

Printed in Singapore

Contents

Contents

Preface

A comparison of *The Principles of Natural Knowledge* with his earlier writings suggests that, among specific ideas, thinking about the idea of *time* was what the physical theory most sharply stimulated in him.

V. Lowe, *Understanding Whitehead*, 1962

Alfred North Whitehead (1861–1947) was a mathematician and logician who, throughout a long scientific career, increasingly questioned the nature of the world around us and the relationships between the fundamental entities that make it up. He became convinced that the concepts transported from everyday life into the philosophy of nature and science: time, space, particles, radiation, and so on, were abstractions and gave us a distorted image of reality. He therefore invented or borrowed from other thinkers a different vocabulary, one that he felt was more suitable for speaking correctly about these things, without being influenced by our everyday habits. The result, of course, has been a rather abstruse discourse and a well-deserved reputation for difficulty. But it is that you have to slowly become imbued with his thinking to be able to empathise[1] with him.

The most important word in his vocabulary is *process*, a word the translations of which into the French language are

---

1 The word 'empathy' is used here in its scientific sense; that is, the ability to put oneself in the place of others in order to understand one's thoughts and follow one's reasoning; it does not imply a value judgement.

misleading. Process is the incessant turnover of nature, the passage from the virtuality of the future to the actuality of the present and its relegation into the past. But it is also the weaving of the web of interactions between the elements of the world, their synergies, their symbioses, their links of cause and effect, which create a world that the philosopher sees as more united than composed of individualities. This is why he calls his work "Philosophy of Organism". In addition, it is an act of creation that is constantly renewed: the unforeseen, the new, it is in the process that these small or large bifurcations that change the world and its course can infiltrate; it is also a creator of time, in the sense that we understand it, because it is in the course of the process that the fundamental entities that make up the world are born and reborn, adding to each birth a parcel of time, imperceptible but finite, that is added to all clocks of nature.

About the Author

Rémy Lestienne, honorary research director at the Centre National de la Recherche Scientifique in Paris, has successively devoted his research to high-energy physics and to neuroscience. From 1998 to 2004, he was President of the International Society for the Study of Time. He is the author of *The Children of Time* and *The Creative Power of Chance* (University of Illinois Press, 1995 and 1998), and *Le Cerveau Cognitif* (CNRS Editions, 2016).

CHAPTER 1

A Brilliant Mathematician

WHITEHEAD'S CHILDHOOD

Alfred North Whitehead was born on 15 February 1861, at Ramsgate, on the easternmost point of the coast extending the south bank of the Thames, closest to Dunkirk. His father, aged 33, was a pastor of the Anglican Church and teacher at the local school, and his mother, then aged 28, ruled a family of four children. She came from a middle-class London family, the Buckmasters. The newcomer's siblings were Charles Selby, 8, Henry, 7, and Shirley, his older sister, then 2.

As detailed in the superb biography of Whitehead written by Victor Lowe[1] (Whitehead's former student and friend, and professor of philosophy at John Hopkins University), from which I borrow a large part of this story, the childhood of the youngest member of the family was a happy childhood filled with parental affection. His father provided his youngest child's first education at the school he ran. Alfred North felt particularly at home in the crisp air of the North Sea, and he had a lifelong attachment to Kent, the region near Ramsgate.

---

1 V. Lowe, *Alfred North Whitehead, the Man and His Work, Volume I: 1861–1910* (Baltimore: The John Hopkins University Press, 1985).

It must have been with sorrow that he left these landscapes when his parents sent him, at the age of 14, to Sherborne High School, 200 miles to the west. It was indeed a school with a growing reputation, and one of Alfred's uncles was a teacher there. Alfred North's parents wanted the best education for him.

In September 1875, Whitehead arrived in Sherborne, a small market town dominated by an old Tudor castle from the sixteenth century. There he received an excellent education in a setting that encouraged freedom, spontaneity, and initiative. His teachers soon noticed his mathematical abilities and exempted him from some homework, such as Latin composition and reading Latin poems to give him more time for mathematics. According to Whitehead himself, his maths teacher in the sixth form was—by far, Whitehead insisted!—the best maths teacher he ever had. Yet he had many others, also excellent.

Not surprisingly, Whitehead won the mathematics medal in 1878, 1879, and 1880, the last year of his schooling at Sherborne. For this final year, moreover, the principal of the school chose Whitehead as head prefect, something of a general supervisor, because he considered him quite capable of leading the boys with authority outside the classrooms. As part of this role, the principal once asked Whitehead to help him suppress homosexuality in the school. Whitehead later told his son that he had been a bit sloppy in this repugnant task. That year of responsibility taught him that authority had to be exercised with restraint: a dose of emotional attention had to be added for authority to be effective, as well as consideration of the natural diversity of personalities.

Naturally, Sherborne was an Anglican religious school, and at that time Whitehead was a fervent believer, in the spirit of the education he had received from his family. But for Sherborne's students, it was natural to read the New Testament in Greek rather than English. Reverence for Greek culture meant that 'we were very religious, but with the natural moderation that befits people who take their religion in Greek'[2] he later recalled.

WHITEHEAD AT CAMBRIDGE

In 1879, while still attending school in Sherborne, Whitehead came to Cambridge during Easter week to take part in an entrance exam, with the possibility of a scholarship. The competition was fierce, and many applicants had to apply for several times. Three £75 scholarships were on offer, and three more of £50, each renewable for three years. Whitehead won one of the £50 scholarships to study mathematics at Cambridge, Trinity College, the temple of mathematics in England.

In October 1880, Whitehead landed at Trinity College, Cambridge. Whitehead was the only new student coming from Sherborne College that year. He was placed, along with about thirty others of his new classmates, under the tutoring of Henry Martyn Taylor, a teacher of geometry who became Whitehead's favourite branch of classical mathematics. Lowe notes: 'accounts of Taylor mention his strict standard of verbal and logical accuracy, and it is likely that Whitehead's insistence on this in his mathematical writings owes something to him. Also, Taylor is known to have imparted enthusiasm for hydro-dynamics, the only branch of applied mathematics on which Whitehead published papers'.[3] The group of mathematics professors included, in addition to Henry Taylor, James Glaisher, a specialist in number theory, and William Niven, who taught Whitehead Maxwell's theory of electromagnetism.

---

2   Ibid., 52.
3   Ibid., 94.

Whitehead was a rather reserved boy and disliked those who used to show off. However, the custom at Cambridge was for students and teachers to get together every night to talk about student life or other subjects. Whitehead complied with this habit, and it allowed him to become acquainted with brilliant students who broadened his curiosity and provided him with his first education in fields other than mathematics. There was, in particular, William Sorley, a young philosopher who later became one of the leaders of British idealism; D'Arcy Thompson, a biologist and future author of *On Growth and Form*, which is thought to have later inspired Whitehead in the development of his philosophy of organism, and Henry Head, future neurologist specialising in somaesthesia (the study of sensitivities inside the body and skin). A little later, when John McTaggart came in 1885 as a student at Cambridge, he was very much involved in Whitehead's training in philosophy. A specialist of Hegel and one of the representatives of British idealism, he would later be the author of an article entitled "The Unreality of Time", based on a critique of the classification of events in the past, present, and future, and remained famous. Perhaps he influenced Whitehead for his particular sensitivity to the question of time.

THE 'APOSTLES'

It may look like a secret society, a figment of J. K. Rowling's imagination, but in fact, it is a venerable two-hundred-year-old Cambridge students' association. Founded in 1820 by twelve founding students (hence its name), the association selects its members by consensus from amongst the best students in the university, from all disciplines, without them applying. Amongst the qualities required, the emphasis is placed on intellectual originality and honesty: the 'Apostles' scrutinise and judge potential young candidates several months before proposing them for election. Upon entry, the laureates must take an oath to keep their membership in the society secret. After a few years of practice, the Apostles take wings and become 'Angels'. They retain the right to participate in meetings, a right they exercise more or less regularly. Much has been said about the Cambridge Apostles' association in the 1950s, when two of its members were prosecuted, accused of passing nuclear secrets to the Soviet authorities.

The Apostles of Whitehead's day met once a week on Saturday evenings in the room of the host chosen by the Apostles for this meeting, over a cup of tea often accompanied by anchovy toasts. It was a very valued friendship and discussion group. The inviting Apostle had prepared a talk on a question of general interest, often with a philosophical component, decided at the previous meeting. His presentation was then followed by a broad discussion, followed by a vote for or against the proposal under discussion, or, quite often, on another more burning issue that the discussion had brought up.

Whitehead was elected Apostle in May 1884, after almost three years of study at Cambridge. His attendance at the club gave him great opportunities to advance his knowledge of philosophical and social issues. His goodwill and lack of selfishness were already evident before he joined the club. But, as Lowe notes, he was far from being Rotary-minded. He had many friends, but no close ones; he was—and was to remain all his life—above all, a lonely spirit. He talked frankly about everything but confided unreservedly to no one.

Amongst the Apostles he frequented a few elders, such as Frederick Maitland, a lawyer and historian; Arthur Verall, a scholar of classical literature; Henry Jackson, specialised in Greek philosophy; and Henry Sidgwick, a philosopher and economist with broad interests. Amongst the younger Apostles, Whitehead felt intellectual sympathy with Goldsworthy

Dickinson, a young Neoplatonist philosopher; in May 1886, Whitehead himself proposed and gained the election of John McTaggart, the philosopher we've already met, one of the future fighters against the reality of time. At the time, he claimed to be a materialist and close to John Stuart Mill, but he quickly specialised in Hegel's philosophy. Very active amongst the Apostles, he had a great influence on Whitehead who, thanks to him, became familiar with this author. Whitehead later recounted that he had been influenced by Hegel without ever having read him, thanks to McTaggart, but that he had read the latter's three books on the idealistic philosopher.

Lowe relates that, in Whitehead's time, 'the majority of the *Apostles* were believers, and Whitehead declared himself so'. At the meeting of 2 May 1885, the question asked was 'Do we believe in God?', and the speaker and moderator was Walter Raleigh. Attendance was twice as large as usual. Maitland, Jim Stephen, and three other members declared themselves atheists. (McTaggart was not yet a member.) Nine declared themselves believers, including Rayleigh, Dickinson, and Whitehead. Two years later, a contribution by McTaggart raised the question, 'Is a personal God a satisfactory explanation of the Universe?' McTaggart and two others answered No; Whitehead and another answered Yes. The young Whitehead believed in God as the creator of heaven and earth, a belief from which he later distinguished himself with particular vividness, if creation is understood as the dated act of a personal God. In 1885, to the question 'Shall we transcend our limitations?' Whitehead wrote on the Society's notebook, 'I want to see God'. The beatific vision was important to him. In October of the same year, he made a contribution on 'Is one life enough?' and the question that was put to the participants at the end of the session was 'Do we desire immortality?' Whitehead answered affirmatively and explained, 'I want a sort of a something'.[4] It should be noted, however, that at the same time, Whitehead was hesitating between the Anglican Church and Catholicism. He even went to visit Cardinal John Henry Newman, probably in 1890, a figure who impressed him greatly. But ten years later, he slipped into agnosticism.

THE TRIPOS OF MATHEMATICS

From the second year onwards, Cambridge mathematics students are immersed in an extremely competitive training cycle, the tripos. This barbaric name has been part of student folklore at Cambridge since immemorial time; it derives from the name of a three-legged stool that was the traditional seat of medieval university professors, when they arbitrated the *disputatios* they proposed to their students. The final examination of the tripos distinguishes the best students, who are given the name *wrangler*.

The difficulty of the competition is such that it does not offer any chance to a student who has not contracted the services of a coach, responsible for familiarising him with the type of tests to which he will be submitted for five days and to respond with the required speed. Whitehead was fortunate to have Professor Edward Routh as his coach, who was himself a *senior wrangler* in 1854 and was considered the best coach of his time. In the course of his career, this professor prepared twenty-seven or twenty-eight first laureates. He gathered his students in groups of up to ten students three times a week.

The atmosphere in the tripos preparation classes is reported in an article published by Andrew Forsyth, who arrived at Cambridge two years before Whitehead and won the

---

4 Ibid., 139. Whitehead clearly means here 'I want something like that'.

competition as a senior wrangler in 1881.[5] 'In my own Tripos in 1881 we were expected to know any lemma in that great work [the Newton's *Principia*] by its number alone, as if it were one of the commandments or the 100th psalm'. In this article, Forsyth, who later reformed the mathematics tripos, is severe on that institution: it was so demanding on the students that, by comparison, the university's courses were passing in the background; other teaching controls were rare, and teachers and examiners changed from year to year, while coaches remained. In short, the institution tended to make students mechanical calculators or computing machines, rather than real engineers or real thinkers.

On 19 June 1883, Whitehead was tied for fourth wrangler. Once relieved of this competition, Whitehead's originality could finally begin to emerge.

WHITEHEAD, NIVEN, THOMSON, AND MAXWELL

James Clerk Maxwell—the brilliant creator of the theory that combines the description of electricity and magnetism into a single theory, soon summarised in four partial differential equations (now known as the Maxwell equations), and which immediately included the description of light as a special case of electromagnetic waves—was a former student and fellow of Trinity College. After a brilliant academic career, he returned in 1871 as a professor of physics at that university. Two years later, he published his book *A Treatise on Electricity and Magnetism* and taught it to some Trinity College students. Lowe writes that 'his professorial lectures had been of little help toward understanding his formidable book, and he was such a poor lecturer

---

5   A. R. Forsyth, "Old Tripos Days at Cambridge," *The Mathematical Gazette* 19 (1935): 162–179. doi:10.2307/3605871.

that very few had gone to hear him—two in 1878–79'.[6] In 1879, the year before Whitehead entered Cambridge, he died prematurely of abdominal cancer. Fortunately, Niven, who had been a student and close friend of Maxwell, took over. He taught Maxwell's theory at Trinity College, and this time his enthusiasm was contagious. Almost all the mathematicians and many of the university's mathematics students attended his lecture in 1884. Whitehead chose this subject for the dissertation he was to write and defend for a research position at Trinity College.

To complete this dissertation and develop his personal thinking, Whitehead also took the course on electromagnetism taught by Joseph Thomson, the great physicist and future discoverer (in 1897) of the electron. Finally, in October 1884, Alfred North Whitehead presented his dissertation. Unfortunately, it is lost today. 'Trinity College did not begin to preserve fellowship dissertations until 1896, and Whitehead did not keep anything for long. … He had no autobiographical inclination, and was the sort of man who has no place in his mind for work that he did when young and passed beyond'.[7]

From a practical point of view, however, Whitehead's dissertation enabled him to obtain a fellowship at Trinity College on 9 October 1884. The circumstances of this success give us a trait of Whitehead's character: he expected so little to be appointed on the first attempt that he left Cambridge as soon as the exams finished and without leaving his address! But at the end of October of that year, he found himself teaching and receiving a scholarship from this institution; the following year was fertile for his training as a mathematician but also as a philosopher, thanks in particular to his admission to the Society of the Apostles.

---

6   Lowe, *Alfred North Whitehead*, Vol. 1, 95.
7   Ibid., 106.

RUSSELL, STUDENT, AND THEN COLLEAGUE OF WHITEHEAD

In October 1889, Whitehead marked the entrance examinations for new students at Trinity College. Amongst the papers was one by Bertrand Russell, eleven years his junior, which particularly caught his attention. When this intriguing new student arrived at Trinity College, contact was established, and Whitehead quickly gained his confidence, especially because, Lowe recounts, ten months after Russell's arrival at Trinity, the teacher reminded the students in his class of a detail from his entrance exam paper, a detail which he had retained.[8]

Russell's initial interest was not in mathematics, but in philosophy. In fact, he did not prepare for the mathematics tripos, but rather for that of moral sciences. He won it brilliantly, in 1894, as senior wrangler. However, he chose it as the subject of his dissertation for his fellowship at Trinity College 'The Epistemological Foundations of Geometry'. He sought the advice of James Ward, a philosopher specialised in Kant and Leibniz, who gave him the advice requested for the philosophical part of his dissertation, but recommended that he go to Whitehead for the mathematical part of his project. Ward thought that the universe is composed of psychic monads, and tended towards pantheism, or rather a philosophical panentheism.[9] Such ideas may be close to those later developed by Whitehead himself, but the two men do not seem to have had much direct contact.

---

8   Ibid., 222.
9   Panentheism differs from Spinoza's Pantheism in that it does not assert, like the latter, that God and the universe are one and the same, but only that the divine is present everywhere and in everything in the universe.

Russell presented his dissertation in 1895. On 9 October of that year, he was received by Ward and Whitehead, both of whom were very critical of him and left him with the depressing impression that it would be failed. But the final verdict came down the next day and was positive, to Russell's complete surprise. From then on, he and Whitehead became colleagues as Trinity College Fellows.

At the end of July 1900, Whitehead and Russell travelled to Paris together to attend two conferences, the first in philosophy (the first International Congress of Philosophy), the other in mathematics (the second International Congress of Mathematics). At the first, Russell read a paper on 'L'idée d'ordre et la position absolue dans l'espace et le temps'. Whitehead did not speak at either congress but served as one of the five secretaries of the second. However, this brief stay in Paris marked a turning point in the careers of both men. Until then, the two colleagues had worked separately. Both had already published important books. For Whitehead, it was *A Treatise on Universal Algebra*, and for Russell *An Essay on the Foundations of Geometry* and *A Critical Exposition of the Philosophy of Leibniz* (a book that Whitehead liked and helped him to get acquainted with the relationist philosopher). Russell was finishing writing *The Principles of Mathematics*, and Whitehead was already working on Volume 2 of the *A Treatise of Universal Algebra*. At the Congress of Mathematics, they met the Italian mathematician Giuseppe Peano, and were seduced by the rigour of his methods in mathematical logic. Little by little, between 1901 and 1902, the two men realised the complementarity of their work and agreed to write and sign together a great treatise that they would call *Principia Mathematica*, and which would make extensive use of Peano's logical notations.

PRINCIPIA MATHEMATICA

The writing of the work was slow and difficult, despite the existence of draft texts already prepared for the projected second volumes of *Universal Algebra* or *The Principles of Mathematics*, and required many difficult discussions between the two authors. In fact, it lasted from 1902 to 1913. In writing the *Principia Mathematica*, Whitehead kept the last hand on all matters relating to mathematical logic, leaving the philosophical part to Russell. Whitehead regarded Russell as a brilliant young logician and considered that his duty as his teacher was to help him. Russell regarded Whitehead as a solid mathematician with a passion for mathematical logic. Indeed, the latter defended in the meantime (in 1905) his doctoral thesis, both on the basis of *Universal Algebra* and on the symbolism of Paeno.

The ambition of the *Principia* is to explain how the whole of mathematics can be developed from the concepts of classes and correlations alone. Whitehead gave a brief analysis of this in "The Organization of Thought", a lecture given in 1916 to the mathematics section of the British Association for the Advancement of Science.[10] In that work, he distinguishes seven kinds of correlations. The most fundamental are those that link one to the multiple, the multiple to one, and one to one. The second are those that arrange the members of a set in a defined order. The third concerns inductive relations, that is, those that depend on a generalisation. The fourth concerns selective relations, on which the construction of arithmetic depends. The fifth, the vectorial relations. The sixth, the relations that link number and quantity, and finally the seventh, the foundation of

---

10   A. N. Whitehead, *The Organization of Thought* (London: William & Norgate, 1917). Lecture reproduced in: *The Interpretation of Science, Selected Essays* (Indianapolis: Bobbs-Merrill Co., 1961).

geometry, concern three- or four-dimensional relations. All of this is developed in the three volumes of *Principia Mathematica*, published from 1910 to 1913, effectively using and developing the Italian master's logical notation system.

In fact, one of the major difficulties in the collaboration between the two authors is their vision of space and the status to be given to *points*. In his book *The Principles of Mathematics*, Russell confesses that Whitehead persuaded him to give up the primacy of points of space and instants of time, a primacy to which he was personally inclined, as the title of his 1901 paper in Paris suggests.[11] But at the end of the work, he makes this correction:

> I conclude, from the above discussion, that absolute position is not logically inadmissible, and that a space composed of points is not self-contradictory. [...] There is no reason, therefore, so far as I am able to perceive, to deny the ultimate and absolute philosophical validity of a theory of geometry which regards space as composed of points, and not as a mere assemblage of relations between non-spatial terms.[12]

We point here a difference in sensitivity between the two authors, which will prove crucial for Whitehead.

Finally, in October 1908, the manuscript of the first, nearly completed volume was handed over to the Cambridge University Press, which considered the work too expensive to print. The publisher asked the authors to apply for a grant from the Royal Society (which they obtained), but also asked for a personal contribution of £50 each. The three volumes of the

---

11   B. Russell, *The Principles of Mathematics*, 2nd ed. (London: Routledge, 1903), xxxvii.
12   Ibid., 461.

work were published between 1910 and 1913. Distribution and acceptance of the work was slow. Mathematicians long considered it to be mainly a book on the philosophy of science. In fact, the book contributed significantly to Russell's reputation, then considered as a science philosopher, but did nothing for Whitehead's.[13]

MEETING WITH EVELYN WADE

Whitehead met Evelyn Wade, his future wife, at the end of the 1890 academic year. She had accompanied cousins of the 29-year-old young man on a visit to Whitehead's parents in Broadstairs, Kent. She was almost five years younger than he was. She had character: her father, an Irish captain, probably with rather strong anti-English sentiments, had chosen to expatriate to France. The family settled in Pont-Aven in France, and Evelyn lived in this country until the age of 17, being brought up by the sisters of a convent in Angers. She lived little with her father and did not get along with her mother. Once in England, she quickly gained her independence by teaching French. The impetuosity of the young girl immediately pleased the young man. She was a determined girl, whereas Alfred Whitehead lacked much confidence in himself. When, shortly after this first meeting, he proposed marriage to her, he was surprised by her acceptance. Later, Evelyn told her daughter Jessie that she accepted the proposal, because she 'never saw a man who needed me more'.[14] The wedding took place on 16 December 1890, at St. Jude's Church in London; the ceremony was presided over by Whitehead's brother Henry with the help of her father, both pastors. Evelyn's mother did not attend.

---

13 Lowe, *Alfred North Whitehead*, Vol. 1, 292.
14 Ibid., 176.

Evelyn must be mentioned because she was of great importance, not only for Alfred Whitehead's emotional life but also for the influence she had on the young man and later on the thinker. For her, says Lowe, 'history was boring, mathematics unintelligible, and no science meant anything to her'.[15] But she gave beauty a primordial meaning, an obviousness, a key to a Platonic world; this had a profound effect on her husband's philosophy.

At the time of their marriage, Alfred was tormented by the religious question. Evelyn was less so. For the first seven years of their marriage, Alfred was absorbed by the question of conversion to Catholicism and had acquired a large library of books on theology.

After 1897, however, he slipped into agnosticism, not to say atheism. Evelyn followed him, and this period lasted twenty-five years. Whitehead's avowed agnosticism was based on a critical analysis of the workings of the Church from its origins (especially St. Paul and his teachings?) to its most recent decisions (the dogma of papal infallibility). But he never departed from a rule of conduct conforming to Christian morality, to which he was very attached. According to his son North, Whitehead's agnosticism or atheism during this period was nonetheless marked by nostalgia; 'it was as if he wanted to be religious and was being defiantly atheistic'.[16] When he rediscovered himself as a theist after 1923, however, he never ceased to be extremely critical of the official churches, their rites and dogmas.

Very social and outgoing, Evelyn nevertheless suffered from time to time from heart or peri-cardiac pain attacks, most probably of psychosomatic origin, but spectacular and disabling.

---

15 Ibid., 177.
16 Ibid., 190.

Throughout their life together, Evelyn's health was a constant concern for Alfred; yet she survived him by fourteen years!

The close collaboration between Whitehead and Russell during the writing of the *Principia* created a great familiarity between their respective couples. Russell had married in 1894, and the couple stayed with the Whiteheads during several extended stays in Cambridge. Russell quickly developed a loving affection for Evelyn, considering that Whitehead neglected her too much. A particularly spectacular attack of Evelyn's angina pectoris, which he and Russell witnessed one evening in 1901 on Russel's return from an outing, deeply moved and upset both men. For Russell, it was the occasion of a kind of brutal conversion, of 'a kind of mystical enlightenment', as Russell himself later recounted. 'I found myself filled with semi-mystical feelings about beauty, with an intense interest in children, and with a desire almost as profound as that of the Buddha to find some philosophy which should make human life endurable'.[17] This was the decisive event in Russell's conversion to pacifism that marked his entire life.

According to Lowe, Whitehead must have quickly realised Russell's feelings for his wife without letting it show and suffered in silence. We have no indication that either Russell's or Evelyn's loyalty was ever questioned, and Whitehead was unaware for a very long time of the financial help Russell provided to the couple during difficult times. After the publication of the *Principia*, however, the intellectual collaboration of the two men, with such different temperaments, became scarce and eventually dried up, but they never ceased to show deep respect for each other.

---

17 B. Russell, *Autobiography* (London: Routledge, 2000), 149.

CHAPTER 2

The Point Is an Abstraction: The Method of Extensive Abstraction

As we have seen, during the writing of *Principia Mathematica* with Bertrand Russell, Whitehead took on the role of the logician–mathematician, vouching for the rigour of the mathematical form to be given to the work. On the other hand, he tried to let his young collaborator pursue and develop the logical-philosophical structure of the work, trusting the author of *A Critical Exposition of the Philosophy of Leibniz*.

Whitehead insisted, however, that the book should take the greatest possible account of the premises he intended to give to the whole construction: to start from immediate perceptions and from them alone, and not from the abstractions that the millenary culture of numbers and geometric figures has deeply inscribed in our thinking. Now, he said, the point is part of these abstractions. What we have immediate perception of are objects that always occupy a finite region of space, not points. In his 1906 article, he refers precisely to Leibniz to justify the refusal to grant points the status of fundamental element: 'Leibniz's theory of the relativity of space ... shows at least that points in space ... must not be taken amongst objective realities'.[1] In the relational theory of space, space is not considered to be the pre-existing framework that matter more or less diversely fills, but the expression of the relations between the elements of matter.

'If there were no creatures, there would be neither time nor place, and consequently no actual space'.[2] Whitehead's article was long and difficult, and thus went virtually unnoticed. Moreover, at that time, the author's adherence to the relational theory of space was not yet definitive. On the one hand, he looked at points as an elaborate abstraction of geometers and considered it essential never to isolate a point or region from its surroundings. But, on the other hand, he is still shaken by Russell's position, reticent as we saw in the previous chapter regarding his master's precautions and warnings. As late as 1910, Whitehead compared the merits of the classical theory of absolute space, composed of points, and relational theory: 'There is to this day no decisive argument for either of these interpretations'.[3] As Victor Lowe notes in this connection, it is therefore necessary to date his definitive adherence to relational theory between 1910 and the beginning of 1914, when he read his address *La théorie relationniste de l'espace* to the congress of *Logique mathématique* in Paris. In this article written in French, points cannot be taken as primary entities for the construction of geometry, as soon as the absolute conception of

---

1   A. N. Whitehead, "On Mathematical Concepts of the Material World," *Philosophical Transactions of the Royal Society, Series A* (Baltimore: Johns Hopkins Press, 1906), 467.

2   Leibniz, 1715–16. G.W. Leibniz and Samuel Clarke, *Correspondence*, Trad. R. Ariew, (Indianapolis: Hackett Publishing Co., 2000), 61.

3   A. N. Whitehead, "The Axioms of Geometry," *Encyclopedia Britannica* 11th ed. (1910).

space is rejected in favour of a relational conception, or as he calls it, relational or relativistic:

> Geometry [...] is accustomed to taking as its starting point all or some of the fundamental spatial entities, points, curved or straight lines, surfaces and volumes. [...] For the relativistic theory of space it is essential that points, for example, be complex entities. [...] For if a point is a simple thing, incapable of being logically defined by means of relations between objects, then points are, in fact, absolute positions. [...] But this is none other than the absolute theory of space which, nominally at least, has been almost universally abandoned. Thus, the first occupation of geometers, seeking the foundations of their science, is to define points through the relations between objects.

> In other words, physical bodies should not be considered as being first in space and then acting on each other; they are in space because they interact, and space is only the expression of certain properties of their interactions.

It is therefore necessary to start from what is immediately given by the senses with regard to space: areas, volumes, permanent objects. As for the relations between two volumes, the most immediate is that of contact: this is the one that allows us to understand how one volume can influence another. Let's not forget that it was in order to free himself from the Newtonian notion of remote action that Faraday introduced the notions of electric and magnetic fields, on which Maxwell built his theory of electromagnetism. An electric charge A does not remotely influence another electric charge B, but creates a field in all space, which being present in location B, exerts a local contact force on the charge there. The notion of contact action, argues Whitehead in his 1914 article, is one more reason to reject the point as a basic principle: because two distinct points are always separated by a distance, which, however infinitesimal, does not really allow recourse to the notion of contact force.

The publication of the *Principia* was not yet complete, when Whitehead already aspired to complete or surpass them. For the exposition of the founding principles of geometry, although based on his conviction that geometry concerns the same world as physics, he could no longer leave aside the evidence of the essentially dynamic nature of the world in which we live. This was a long-term program that he, henceforth, intended to work on. In the following years, between 1914 and 1920 (the year of the publication of *The Concept of Nature*), Whitehead first refines the arguments for rejecting the point as a fundamental existing in the description of nature and consolidates the theory of *extensive abstraction*, a mathematical theory he developed to construct the abstract point and relate it to the physical quantities that can be associated with it in physical theory.

In *The Organization of Thought* (1916), Whitehead concisely states the reasons for the program he intends to develop. In the classical conception of physics, he says, 'we imagine that we have the immediate experience of a world of perfectly defined objects implicated in perfectly defined events which, as known to us by the direct deliverance of our senses, happen at exact instants of time, in a space formed by exact points, without parts and without magnitude. [...] My contention is that this world is a world of ideas, and that its internal relations are relations between abstract concepts'.[4]

In "The Anatomy of Some Scientific Ideas" (1917), he explains: 'The space-relations between the parts are confused

---

4   A. N. Whitehead, *The Organization of Thought* (London: William & Norgate, 1917). Lecture reproduced in: *The Interpretation of Science* (Indianapolis: Bobbs-Merrill Co., 1961), 19.

and fluctuating ... The master-key by which we confine our attention to such parts as possess mutual relations sufficiently simple for our intellects to consider is the principle of convergence to simplicity with diminution of the extent. We will call it the "principle of convergence".[5] The rest of the article will be of particular interest to us, because even before applying this principle of convergence to points in space, he applies it to time.

> The first application of the principle occurs in respect to time. The shorter the stretch of time, the simpler are the aspects of the sense-presentation contained with it. The perplexing effects of change are diminished and in many cases can be neglected. Nature has restricted the acts of thought which endeavour to realise the content of the present, to stretches of time sufficiently short to secure this static simplicity over the greater part of the sense-stream.

How can we represent the process of abstraction of the point of space, extended from 1911 to the instant of time, as we will see in Chapter 4? The simplest image is to consider two spheres or on the plane two concentric circles of visible size. Let us declare the area between the two circles to be 'real' or 'concrete', excluding what is in the small circle; let us apply the principle of convergence to this figure, reducing this image by homothety: the two circles remain contained within each other and the surface between them is still measurable, but the limit point, the one that Euclid defined as 'that of which there is no part', remains an abstraction devoid of reality (Figure 2.1).

---

5   A. N. Whitehead, "The Anatomy of Some Scientific Ideas."
     (New York: Macmillan, 1917). Lecture reproduced in: *The Aims of Education and Other Essays* (New York: Macmillan, 1929), 128.

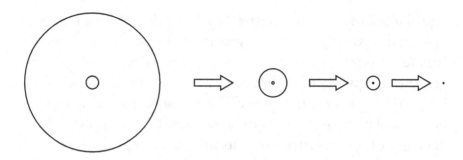

Figure 2.1. *The geometrical point is an abstraction. Let us declare as real the space between the two concentric circles, excluding the space within the small circle—if only because it is too small to be discerned (but for other reasons as well); an abstract point (having no concrete existence) can be constructed by reducing this image indefinitely.*

Let us now think of this process applied not to circles but to spheres in the space around us. In this process, the volume considered real and concrete (between the two spheres) does not converge towards anything real. All volumes become indefinitely smaller and smaller as thought progresses in the series of increasingly smaller spheres; but there is no absolute minimum that is finally reached. On the other hand, it should be noted that the physical properties of the volume under consideration generally tend towards a finite value (e.g. temperature, or electric field). These finite values characterise, in fact, not properties of the 'point', but relations between parts, between the point and its environment, which explains why they can be real when the point itself is only an abstraction. This remark thus validates the relational point of view of space that Whitehead wants to defend. Finally, these values define the physical state 'at this point' and as real quantities they indeed allow the development of a physical theory.[6]

---

6   In particular, it is worth considering that the 'field' as defined by Maxwell at a point requires a 'vanishing test charge' producing no disturbance at that point.

THE METHOD OF EXTENSIVE ABSTRACTION IN *THE CONCEPT OF NATURE*

Whitehead refined the mathematical treatment of the abstract point and detailed the chosen method, entitled the *Method of Extensive Abstraction*, in several writings. We will retain in particular the one presented in *The Concept of Nature*, intended for a large audience.

The fundamental entity that we will retain is that of event. It is a very general concept: something happened in a certain region of space, or nothing in particular happened there, but we paid attention to it. For Whitehead, it's at once a four-dimensional volume, because an event, as a general rule, happened in a region of space for a period of time. Whitehead geometer has no particular difficulty in adopting this relativistic viewpoint, perhaps borrowed from Minkowski.[7]

Second, the fundamental relationship between events, on which the theory will be based, will be that of overlap, or according to the terminology adopted by Whitehead, of *extension*. 'If an event A extends over an event B, then B is a "part" of A, and A is a "whole" of which B is a "part".'[8] It is easy, using two-dimensional diagrams of events A and B, to see that between two events four cases must be considered: either A overlaps with B, or B overlaps with A, or A and B have a common part, or A and B are totally disjoint. Furthermore, let us note with Whitehead that 'if A and B are two events, and A' a part of A, and B' a part of B, then in many respects the relations between the parts A'

---

7   H. Minkowski, Talk 'Space and Time' delivered in 1908 in Cologne. English Translation: 'Space and Time', 1909, https://en.wikisource.org /wiki/Translation:Space_and_Time. See also hereafter, Chapters 4 and 8.

8   A. N. Whitehead, *The Concept of Nature* (Cambridge: Cambridge University Press, 1920), 76.

and B' will be simpler than the relations between A and B. This is the principle which presides over all attempts at exact observation.'[9] This justifies the convergence method which, as we have seen, makes it possible to define not only the point but also other geometric entities such as the line or the plane.

From now on, this will be referred to as an abstract set of events. It is a set of events that has the following two essential properties: (*a*) *between any two members of the set, one contains the other as its part* and (*b*) *there is no event that is a common part of each member of the set.* A little reflection will easily convince you that this second condition rejects the geometrical point as not being part of the abstract set under consideration. The point in four-dimensional space is not an event.

To familiarise ourselves a little with this abstract ensemble concept, let us consider the event patterns presented in Figure 2.2. Let's say that in these diagrams, the spatial dimensions are compacted into one dimension, horizontal, and time is represented by the vertical dimension. In Figure 2.2A, two abstract sets are represented, the set of squares and the set of circles within those squares. It is clear that, when we go through these two sets in order of decreasing dimensions, both converge towards point P (although this is not part of the abstract set). Because they both converge towards the same limit, we will say that the two sets are equal. Now let's consider scheme B, that of the circles and that of the rectangles contained in the square inscribed in the larger circle, when we decrease their height. These two sets do not converge in the same way: that of the circles still converges towards the point *P*, but that of the rectangles towards the line segment *l*. In four-dimensional space–time, the generalisations of these types of abstract sets make it possible to define points, lines, planes, or volumes. In order not to confuse them with the entities defined in classical geometry of

9   Ibid., 79.

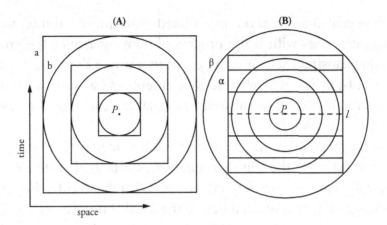

Figure 2.2. *In (A), the two abstract sets a and b, consisting of smaller and smaller squares or smaller and smaller inscribed circles, respectively, both converge at point P. In (B), the two abstract sets a and β, consisting of increasingly narrow rectangles or increasingly small circles, converge the first to the line segment l, and the second to point P.[10]*

absolute space, Whitehead calls them event-particle, point trace, and matrix, respectively, because it is appropriate 'to reserve the ordinary spatial terms "plane", "straight line", "point", for the elements of the timeless space of a time-system.[11] Let us retain for the moment the term event-particle, but let us forget the other two, which we will not need. Let us note that, strictly speaking, for Whitehead, an event-particle is still an abstract element forming part of an abstract set, represented by all the abstract sets that tend towards it, but without going as far as the dimensionless geometric point, excluded from reality. The totality of the event-particles, for their part, form the four-dimensional space–time, the fourth dimension emerging from time. Thus, the notion

10 These diagrams are based on R. M. Palter, *Whitehead's Philosophy of Science* (Chicago: The University Chicago Press, 1960), 43 and 51.
11 Ibid., 94.

of the event-particle is not just a fixed point in four-dimensional space; it carries with it the process of convergence of the group of 'equal' abstract sets. It is therefore, in a way, a dynamic notion: 'an event-particle is an abstractive element, and as such is a group of abstractive sets those abstract sets that are each equal to some given abstract set'.[12]

It has been seen that, although geometric points are not part of abstract sets, the physical quantities attached to these sets generally converge to defined values. Whitehead calls the set of convergent quantities that define the local state of a system, but which are related to the relationship between the whole and the parts, the intrinsic character of the extensive set in question. This remark is important, because Whitehead never wants to forget that an event-particle is never more than the focus of an abstract whole, 'nourished' by relations with its environment. Cut off from these relationships, the event-particle cannot exist:

> An isolated event is not an event, because every event is a factor in a larger whole and is significant of that whole. There can be no time apart from space; and no space apart from time; and no space and no time apart from the passage of the events of nature. The isolation of an entity in thought, when we think of it as pure 'it', has no counterpart in any corresponding isolation in nature. Such isolation is merely part of the procedure of intellectual knowledge.[13]

Let us now turn our attention to a particular abstract set: that of horizontal rectangles as in Figure 2.2B; let us consider a small rectangle, such as the one facing the letter α and build from it a new abstract set, composed of rectangles such as this one, but whose large sides are progressively enlarged. Thus,

---

12 Ibid., 89.
13 Whitehead, *The Concept of Nature*, 144.

the small sides of this rectangle are finally rejected *ad infini-tum*. This type of abstract set tends towards an unlimited slice of space between two times, represented on the diagram by the lines extending the two large sides of the initial rectangle. Whitehead calls such a slice of space–time a *duration*.

Let us now continue this game by defining the abstract set obtained by progressively bringing the two preceding lines closer together, thus compressing the duration towards an instant. The slice of space included in the duration converges towards a 'line' or a 'punctual trace' of infinitely narrow tempo-ral thickness.

Whitehead calls such an abstract element a *moment*. A moment thus represents the whole of nature at a given time.

Whitehead has the audacity to write in *The Concept of Nature*:

> Each moment is a group of abstractive sets, and the events which are members of these sets are all members of one fam-ily of durations. The moments of one family form a tempo-ral series; and, allowing the existence of different families of moments, there will be alternative temporal series in nature. Thus, the method of extensive abstraction explains the origin of temporal series in terms of the immediate facts of expe-rience and at the same time allows for the existence of the alternative temporal series which are demanded by the mod-ern theory of electromagnetic relativity.[14]

This is an opportunity for him to open his mind to Einstein's theory of relativity. In four-dimensional space–time, the divi-sion between space and time can be conceived in different ways, for example, by rotating the axes representing the differ-ent coordinates of a given frame of reference.

---

14 Ibid., 89.

MEASUREMENTS IN SPACE–TIME AND THE CONGRUENCE PROBLEM

Now we have a space–time continuum, whose stuff is given in terms of event-particles. To use it in physics, we still need to specify the type of measurement we can make in this continuum. Einstein tends, Whitehead reproached, to assume from the outset that the problem is solved. He endowed his space–time with a generally orthogonal coordinate system; space measurements are performed using rulers and time measurements using a clock. But how do we know that the ruler can be carried everywhere in space–time? How do we know that the clock will beat the same second if it is moved? That's the problem of *congruence*. Concerning the use of pendulum clocks on earth, the idea that the clock is accurate in any displacement is obviously wrong, if only because of the variation of the g-factor (acceleration of gravity) between the poles and the equator. Whitehead asks: how do we know that a minute today is the same as a minute yesterday? Einstein appeals to clocks, but since clocks work differently in different coordinate systems, there is no absolute standard. Whitehead, for his part, appeals to the basic principle of 'recognition'.[15] Recognition, he says, is an immediate factor of consciousness, which occurs in the immediate present without the intervention of memory. We know when two things are the same, and this basic fact is required before we use clocks to measure time. That is why we are constantly looking for a clock that, given our knowledge, is as accurate as possible for travel in time and space. This is obviously what technology is striving for, moving away from mechanical clocks to atomic clocks.

---

15 Ibid., 130.

PRIMACY OF TIME MEASUREMENTS

Whitehead's reflections on congruence led him to observe that measurements of space are conditioned by measurements of time; in short, that time always underlies any measurement operation. In his philosophy, this is an obvious fact, since the true element of reality for him is *process*, and nothing exists except in a duration. Doesn't the measurement of a length, for example, require the displacement of a ruler in time? This remains true, and even more obvious, for geometers or astronomers who measure distances using the propagation of light rays, going to the extent of using light-years as units of distance in astronomy. Of course, this is not an original position. I have already had occasion to point it out,[16] in experimental child psychology, primitively, there is only time, or more precisely, only waiting: anxiety before the desired oral contact, the conscious interval between need and satisfaction generate the notion of distance. Whitehead's vision of his space–time continuum confirms this: it, as we have shown, is made up of event-particles whose dynamic character is underlined by the thought process that traps them, that is to say, by the abstract sets that alone construct them: 'The dependence of the character of space on the character of time constitutes an explanation in the sense in which science seeks to explain.'[17]

In mathematics, the technique of abstractive sets is still today an important conquest of this discipline. In Whitehead's biography published by Gary L. Herstein (*Internet Encyclopedia of Philosophy*), the latter mathematician–philosopher notes that 'extensive abstraction is considered to be foundational in

---

16 R. Lestienne, "Characteristics of Physical Duration," *Scientia* 107 (1972): 76–100.
17 Whitehead, *The Concept of Nature*, 98.

the contemporary field of formal spatial relations known as mereotopology', a theory describing the spatial relationships between sets, parts, parts of parts, and the boundaries between these parts. It is also opportune to question, with another philosopher of science, Thomas Mueller, what becomes, in a physics described by means of extensive abstraction, Laplacian determinism. He observes that physical systems are often in situations close to what is called mathematical chaos: initial conditions that are as close as one wants but slightly different quickly lead to very large discrepancies in the state of the system, 'in general, it appears that the choice of an ontology based on regions re-establishes reductionism, but that it represents a major problem for determinism'.[18] We will see further on that determinism is indeed relativised in Whitehead, not only for this reason but also for more fundamental reasons, because of the mechanism proper to the *process*, which breaks the perfect continuity of nature's time, as opposed to the time of abstract concepts and objects of thought, for which continuity can be maintained.

18 T. Mueller, "The Method of Extensive Abstraction in Physics. The Determinism without Euclidean Points," *Chromatika*, 2009, accessed November 17, 2021, https://serval.unil.ch/resource/serval:BIB_73F28BE9E250.P001/REF.

CHAPTER 3

Whitehead in London: Philosophical Encounters

During his stay in Cambridge, Whitehead's participation in the Apostles' Discussion Circle and the many opportunities to meet with his colleagues at Cambridge and Oxford had already given him a certain familiarity with the philosophical thought of ancient authors, such as Aristotle and Plato, or modern authors, such as Leibniz (1694) and Kant (1781[1]). Proving his interest in philosophy, he had also joined another philosophical think-tank, led by his colleague John McTaggart, the *Eranos Club* (a Greek word that evokes a meal taken in common, where everyone brings their own food). But during all these years, his main pre-occupation was of course the foundations of logic and mathematics. His resignation and his forced stay in London, from 1910 to 1924, could have cut short his career as a logician–metaphysicist. On the contrary, it was a period of intense reflection and exchange with new philosophical minds that would nourish his vision of the world, heralding the metaphysical system he would develop in America after 1924.

---

1   The dates in brackets are dates of publication of major works by the authors concerned.

During his long stay in London, Whitehead gave mathematical contributions that were increasingly marked by philosophy, first in the form of lectures, particularly in Cambridge and before the British Association for the Advancement of Science (BAAS) or the Aristotelian Society for the Systematic Study of Philosophy (AS), later published and finally completed in 1920 by a book, *The Concept of Nature*, written for—he hoped—a wider audience of physicists, natural scientists, and philosophers of science. In the meantime, he perfected his philosophical culture by becoming familiar with the thoughts of authors, such as Locke (1663), Spinoza (1670), Leibniz (1694), Hume (1748), Kant (1781), James (1890), as well as by frequenting living philosophers, such as Alexander, Bergson, Bradley, Broad, Moore, Loyd Morgan, McTaggart, Russell, and Wittgenstein.

AN EVENTFUL RESIGNATION

The departure of the Whiteheads from Cambridge to London in the summer of 1910 occurred in a hurry. It was triggered by a conflict between Whitehead and the administration of the institution over a fellow mathematician and friend, Andrew Forsyth.

Andrew Forsyth, who was his senior by three years, had entered Cambridge two years before Whitehead, and quickly proved to be an outstanding mathematician (he was the senior wrangler of the year 1882). Whitehead immediately admired him and befriended him. In fact, Forsyth's career advancement progressed ahead of his own! As a sign of Whitehead's special consideration for him, he chose him as godfather to his first child, Thomas. Another sign of consideration and friendship, when, in 1906, Forsyth fell seriously ill for several months, Evelyn, Whitehead's wife, devoted herself very much to him. In return, Forsyth facilitated Whitehead's career in England on several occasions.

In 1910, Forsyth, 52 years old and until then known as a perfect gentleman, fell in love with the wife of a colleague, the physicist Charles Vernon Boys. Maria Amelia Boys was the mother of two children, but felt neglected by her husband; she gave in to Forsyth's advances, and soon ran away with him. The affair became public, and the University Council urged Forsyth to resign his office and his fellowship. But Whitehead, who sat on the board, felt that the fellowship could not be taken away from him and slammed the door when the board approved the resignation. Then, two months later, he sent his own letter of resignation. The council then noted that Whitehead had completed the twenty-five years of service to which the university had committed itself since October 1909. From then on, the resignation was automatically accepted; but contrary to the sanction which had been imposed on Forsyth, Whitehead was able to keep his paid title of assistant.

If Forsyth's dismissal had been the immediate factor in his decision, Whitehead had been in the mood for a change of scenery for some time. Cambridge, he thought, did not recognise his true worth. Evelyn, on the other hand, did not enjoy much the company of Cambridge's women professors; she did not feel comfortable in that environment. Whitehead and his wife therefore agreed to try something else. When Victor Lowe later asked Whitehead why he left Cambridge, he replied, 'I was in a rut at Cambridge.'[2]

In the summer of 1910, the Whiteheads moved to London. Evelyn found an apartment to their liking in Chelsea, 17 Carlyle Square. A comfortable bourgeois house in a very popular area at that time. However, Whitehead, who had not prepared his professional future, found himself unemployed (keeping only his paid

---

2   V. Lowe, *Alfred North Whitehead, the Man and His Work, Volume I: 1861–1910* (Baltimore: The John Hopkins University Press, 1985), 315–17.

job as an Assistant in Cambridge). Unemployed, but not without activity, he took advantage of this break to write *An Introduction to Mathematics*, a little-read but remarkable student book, published by Cambridge University Press in 1911 as the 56th volume of their 'Great Books of the Western World' collection.

Fortunately, Whitehead's unemployment ended in July 1911. It so happened that Francis Galton, Darwin's cousin who had been knighted by King Edward VII, left a bequest in 1911 to set up a department of applied statistics and eugenics at the University of London. His disciple, the statistician Karl Pearson, was to take up this chair, but to accept it, he had to give up the chair he had held for 27 years, that of Applied Mathematics and Mechanics at University College. He resigned in June 1911. Thus, the position became vacant, and Pearson suggested to the Provost of University College London that Whitehead be invited to take it. The salary associated with the position was meagre, but Whitehead had no choice and accepted.

This situation did not last. In March 1912, the Faculty Board selected the assistant professor in the Department of Pure Mathematics to succeed Pearson. The assistant professor's position became vacant. Whitehead wrote to the Provost, reminding him of the assurance he had been given that the temporary position he had accepted would not prevent him from applying for a permanent position. As a result, in fact, a new position of associate professor of geometry was created specifically for Whitehead. Two years later, on 1 September 1914, Whitehead was appointed professor of applied mathematics at the prestigious institution of Imperial College, where he continued to work until 1924.

WORLD WAR I AND THE RUSSELL DISPUTE

As the year 1914 progressed, Whitehead became concerned about the international situation and came to believe that France and Germany would clash. He also realised that England could not stand aside. Russell, who was still his best friend at the time, disagreed completely.[3] His pacifism and mistrust of the future Russian ally dictated his position, and he remained inflexible to the exhortations of Alfred and to those of Evelyn who had been brought up in France in a fiercely Germanophobic atmosphere. Their son Thomas North, then 23 years old, enlisted himself as soon as war was declared, with the active support of his parents and his uncle Henry. A year later, in the summer of 1915, North, who had been sent back from Flanders after a strong trauma apparently caused by a shell burst, was sent to fight in Kenya. There, North suffered three attacks of malaria and several attacks of dysentery, and was finally sent back to England in the spring of 1918.

In 1917, the Whiteheads' second son, Eric, then 19 years old, left school to train as a pilot in the Royal Flying Corps. In February 1918, he was sent to fight in France, but on March 13, 1918, his plane crashed in action in the forest of Gobain, in the Aisne.[4] Eric was his mother's favourite son. North and Jessie were known to disobey at times, but not Eric. The shock to the Whiteheads was devastating. For Alfred, it was more than doubled by his empathy for Evelyn's immense grief.

In the meantime, the Whiteheads continued to tolerate Bertrand Russell's militant pacifism, showing him their friendship (while matching their messages with anti-pacifism pleas). At the end of 1915, the latter had taken a stand against conscription. In 1916, as an active campaigner in an association supporting conscientious objectors who had defended them in a leaflet, he took the initiative of writing a letter to the *Times*

---

3 V. Lowe, *Alfred North Whitehead, the Man and His Work, Volume II: 1910–1947* (Baltimore: John Hopkins University Press, 1990), 27.
4 Ibid., 34.

denouncing himself as the author of the leaflet in question. The government decided to take him to court. He defended himself, made a superb defence (which Evelyn attended), was fined £100 which was collected and paid.[5]

But Russell's turbulent actions eventually cooled his relationship with the Whiteheads. In contrast to Russell's attitude in the darkest days of the war, Whitehead continued to display a flamboyant patriotism, as he did in a prize-giving speech in February 1917.

Moreover, the distance between the two men was not only politically motivated. As early as 1914, Alfred was upset to find that *Bertie* (as the Whiteheads called Bertrand Russell) had used working notes he had given him in the book *Our Knowledge of the External World as a Field for Scientific Method in Philosophy*, which he had published that year. On January 8, 1917, Whitehead sent Bertie a stern warning: 'I don't want my ideas propagated at present either under my name or anybody else's [...] I do not want you to have my notes which in chapters are lucid, to precipitate them into what I should consider as a series of half-truths'.[6] However, two days later, Evelyn in turn sent a letter to Bertie, no doubt partly to soften her husband's letter, under the pretext of reproaching him for having neglected her when she was ill. A year later, on 9 February 1918, after Russell was sentenced to prison for his support of conscientious objectors and against the war, Evelyn wrote to him: 'However passionately we may disagree with your present views, to us, you are you, the friend we value, whose affection we count on, the friend whom our boys love, and in many ways still our Infant Prodigy'. Finally, reports Guillaume Durand,

---

5   Ibid., 36.
6   B. Russell, *The Autobiography of Bertrand Russell* (New York: Bentham Book, 1967–1969), 96.

the differences in epistemological options between the two thinkers continued to deepen; if, in 1914, Russell still wrote: 'the instants are not among our data of experience [...] we must build them',[7] he would not be long in criticising the method of extensive abstraction and departing from Whitehead's fundamental choices, as we saw in the previous chapter. In 1927, in *The Analysis of Matter*, Russell severely judged the approach of his former master: 'Dr. Whitehead assumes that every event encloses and is enclosed in other events. There is, therefore, for him, no lower limit or minimum, and no upper limit or maximum, to the size of events [...]. For the above reasons, I am unable to accept Dr. Whitehead's construction of points'.[8]

THE ARISTOTELIAN SOCIETY AND THE ENGLISH EMERGENTIST MOVEMENT

The war years were also an opportunity for Whitehead to engage in intense philosophical reflection and exchange with professional philosophers. Whitehead joined the Aristotelian Society of London in 1915. There he met Bertrand Russell, a member since 1896. He also frequented eminent personalities such as Thomas Percy Nunn, D'Arcy Thompson (his former fellow student at Trinity College), and Richard Burton Haldane, future Lord Chancellor and influential philosopher. But the most important personalities from the point of view of philosophical exchanges were undoubtedly Conwy Loyd Morgan, biologist, Charlie Dunbar Broad, philosopher, and Samuel

---

7   G. Durand, "Whitehead et Russell: la discorde de 1917," *Noesis* 13 (2008): 237–50.
8   B. Russell, *The Analysis of Matter* (London: Kegan Paul, Trench, Trubner, 1927), New edition (Nottingham: Spokesman, 2007), 292–94.

Alexander, also a philosopher, and all three founders of the English emergentist movement. Emergentism in science is essentially characterised by one observation and one conjecture: on the one hand, we observe that in nature we are witnessing a stratification in levels of increasing complexity, each level requiring an almost autonomous description, although interactions between levels are also present. For example, one observes the levels of elementary particles, atoms, molecules, and so on, or those of viruses, cellular beings, organs, living beings without nervous system, evolved animals, the human being possibly distinguished by his consciousness. Second, emergentists believe in the conjecture that, when some of these steps are taken, new, so-called emergent properties appear, sometimes explained by the appearance of actions of the whole on the part ('The whole is more than the sum of the parts' is an aphorism attributed to Aristotle) and properties that in any case could not be predicted by the laws that science uses to describe components at a lower level of complexity. Life and consciousness are often cited as candidates for these emergent properties.[9]

When in 1916 Whitehead gave his first lecture to the Aristotelian Society, entitled "Space, Time and Relativity", he presented himself as an 'amateur' in philosophy of science, not out of lack of confidence in his work but rather, according to Lowe, as a mark of professional courtesy. His philosophical reflection was already quite mature, and moreover, was quite in tune with the dominant neo-realism of the time.[10]

In the *Proceedings* of the Aristotelian Society between 1910 and 1924, Whitehead's London period, Russell gave four

---

9   On this subject, see R. Lestienne, *Dialogues on Emergence* (Paris: Le Pommier, 2012). English translation in: "Dialogues About Emergence," *Kronoscope* 16 (2016): 15–135.
10 Lowe, *Alfred North Whitehead*, Vol. II, 100.

lectures, Charlie Broad five, Samuel Alexander one, Conwy Morgan two, and Whitehead five, these mainly in relation to the theory of relativity.

The interactions between Whitehead and Alexander are crucial. Alexander, two years younger than Whitehead, studied philosophy at Oxford; later he studied psychology in Freiburg, Germany. An admirer of Spinoza, he evolved from idealism to realistic epistemology. His main work remains *Space, Time and Deity*, published in 1920. The two authors share common points: both repudiate the idea of duality in nature; both use from the outset space–time, where space and time are indissolubly mixed; and this space–time is the place of perpetual movement, transformation, and evolution. While Whitehead is a relationalist,[11] Alexander absolutises space–time, calling it the fabric of the universe; moreover, Alexander's world is essentially continuous; he would have had great difficulty with quantum mechanics. He does not use the method of abstract extension, albeit knowing it well.[12] Finally, Alexander remains very Kantian in his insistence on the trilogy *Substance, Cause, and Relation*, while Whitehead remains, as we have seen, closer to Leibniz and Spinoza.

Whitehead and Alexander expressed mutual sympathy and congratulations to each other. In May 1924, the former wrote to the latter to tell him of his joy at having made the transition from mathematics to philosophy.[13] The following year, in the preface to *Science and the Modern World*, Whitehead said he

---

11 D. Emmet, "Whitehead and Alexander," *Process Studies* 21 (1992): 137–48.

12 Alexander's preface to the 1927 edition of *Space, Time and Deity* mentions that he knew two of the Whitehead's books: *An Enquiry Concerning the Principles of Natural Knowledge* and *The Concept of Nature*.

13 Lowe, *Alfred North Whitehead*, Vol. II, 153.

was especially indebted to Alexander and his great book *Space Time and Deity* (he also mentions Conwy Lloyd Morgan's book, *Emergent Evolution*). In the last years of their lives, Alexander used to say that in his opinion Whitehead had surpassed him,[14] and Whitehead confided that Alexander was the philosopher of his time who had taught him the most. He added that they both understood the problem of metaphysics in the same way, that is, as the reconciliation of the unity of the universe (a thought dear to Spinoza) with the multitude of individualities (a thought more in accordance with Leibniz).[15]

Given this climate, it is interesting to note the similarities and differences between Alexander and Whitehead on the subject of Time. In terms of similarities, it is worth noting the primary importance given by both thinkers to this notion. While Whitehead, from 1920 onwards, declares that he no longer uses this term and replaces it with *process*, Alexander associates it with *motion*, and adds that time is not only the change of the world, but an inner quality present in all things in the world. He will say that 'Time is the Spirit of Space', a remark that will earn him the sarcasm of the positivist philosophers of his time. During the Harvard Conference about Time that he gave in 1926, Whitehead began by noting that Alexander had asked that time be taken seriously. And he adds: 'If time be taken seriously, no concrete entity can change. It can only be superseded'.[16] It should be noted, however, that Alexander's space–time remains absolutely continuous and composed of 'points-instants', where Whitehead speaks of 'event-particle', that is, the

---

14 Emmet, "Whitehead and Alexander."
15 Lowe, *Alfred North Whitehead*, Vol. II, 173.
16 A. N. Whitehead, *The Interpretation of Science, Selected Essays* (Indianapolis: Bobbs-Merrill Co., 1961), 240.

fruit of a family of abstract extensions denying existence to both points and instants.

The traces of reciprocal influences between Whitehead and the other creators of English emergentism are more diffuse. In *The Concept of Nature*, our author notes: 'Since the publication of *An Enquiry Concerning the Principles of Natural Knowledge* I have had the advantage of reading Mr C. D. Broad's *Perception, Physics, and Reality*.[17] This valuable book has assisted me in my discussion in Chapter II [Theories of Bifurcation in Nature]'. In *Science and the Modern World*, he thanks Lloyd Morgan for his book *Emergent Evolution*, which he found 'very suggestive', but later he will specify that this book, on the whole, did not really strike him. However, it is in this book that for the first time the term emergence is used in its modern sense: 'the orderly sequence [of events in nature], historically viewed, appears to present, from time to time, something genuinely new. Under what I here call emergent evolution stress is laid on this incoming of the new'.[18] The philosophy he develops in this work resembles in many ways that of Whitehead: refusal of the Cartesian duality, an attempt to develop an emergent theory of mind. He repeatedly quotes the *Concept of Nature*, published three years earlier. But he also distances himself from Whitehead for his attempt to characterise nature as a closed system independent of the knowing observer.[19]

On the other hand, we have the reviews of Whitehead's books written by C.D. Broad, published in *The Hibbert Journal*. Commenting on *The Concept of Nature*, Broad writes: 'The

---

17 C. D. Broad, *Perception, Physics, and Reality* (Cambridge: Cambridge University Press, 1914).

18 C. Lloyd Morgan, *Emergent Evolution* (London: William & Norgate, 1923), 1.

19 See A. Gare, "Process Philosophy and the Emergent Theory of Mind: Whitehead, Lloyd Morgan and Shelling," *Concrescence* 2 (2002): 1–12.

notion of durations is fundamental for Whitehead and is very difficult to grasp. [...] I am not perfectly clear whether Whitehead regards the particular immediacy which belongs to durations that do fall within specious presents as a fact of external nature [per se], or as a "psychic addition". [...] Durations are said to be wholes all of whose parts are simultaneous. This sense of simultaneity does not imply "instantaneousness". [...] Simultaneity, in this sense, is [thus] an irreducible *three*-term relation. The events A and B are not, by themselves, simultaneous; you can only say that they are simultaneous with respect to some duration C ...' Further on, Broad continues: 'Time is within nature in the sense that the measurable time of physics expresses the relationships between durations, and durations are slices of nature. On the other hand, time extends beyond nature, in the sense that our mental acts succeed each other; what was perceived ceases to have immediacy and becomes merely remembered or quite forgotten. But the time in which the mind is cannot be identified with the time-series of nature, because mental events do not have those properties of natural durations which lead to a definition of physical time'. These nuanced and complex observations end with a remark apparently addressed to the Cambridge University Press, publisher of *The Concept of Nature*, but indirectly critical of the author himself, who is suspected of negligence: 'The thanks rendered in the Preface by Professor Whitehead to the Cambridge University Press officials seem to me excessive. [...] they have passed at least six bad [typographical] mistakes'.

In Broad's 1947 obituary for Whitehead, Broad notes that in 1945 Whitehead received the highest possible honour in England, the Order of Merit, after having been admitted to the Royal Society in 1903 and having combined this distinction in 1931 with the membership in the British Academy (an almost unique conjunction in the history of England). He attributes to Whitehead a series of scientific merits, including that of having deduced Lorentz's transformation equations in the theory of special relativity from extremely general considerations, 'without reference to such concrete and contingent matters as the synchronisation of clocks by means of light signals'. Indeed, these laws of transformation derive from the consideration of rotations of coordinate axes in space–time; without explicit reference to light, they are established by introducing a constant, only assimilated *a posteriori* to the speed of light in vacuum. Nevertheless, Broad cannot help observing that in his opinion 'he was abominably obscure and careless writer, and this fault certainly grew on him as he became older. It is the more deplorable, since he certainly possessed at one time the power to write clearly, as his little book on *Mathematics* shows, and since he certainly retained that power when he chose to exercise it, as is shown by many passages in such later works as *Science and the Modern World* and *Adventures of Ideas*'. No wonder, then, that a few lines later he describes *Process and Reality* as 'one of the most difficult philosophical books that exists', although he concedes that there is almost certainly 'something important concealed beneath the portentous verbiage of the Gifford Lectures'.

But can Whitehead be classified as an emergentist thinker? Let us recall that for emergentism, nature is an interlocking of systems of increasing complexity, a stack of successive levels, from the atom to the organic being. And the passage from one level to another can foil reductionism, because in these passages, new, unexpected properties may appear that are inexplicable on the basis of the properties of the constituent elements of the lower level.

Whitehead was certainly influenced by the English emergentist movement, at a time when his metaphysics of the world
was beginning to mature within him. But can he be called an
emergentist himself? The answer to this question depends on
what is meant by emergence. British emergentists insisted on
the idea that the emergence of new properties and the creation
of novelty in a system only occurs when thresholds of complexity are crossed. Matter, they observed, is organised into
hierarchical levels, from elementary particles to atoms, molecules, cells, living organisms, thinking beings. These levels were
crossed over billions of years of evolution, and then the individual units slowly organised themselves into societies.

In order to move from one level to another, the system or
organism must have acquired a sufficient level of complexity to
allow this passage, for example, when an inanimate system made
up of levels of chemical interactions organises itself sufficiently
to generate the metabolic reactions that support sustainability,
reproduction, and action that we call life, or the manipulation
of biological information that we call thought. In contrast,
Whitehead says that the possibility of creating something new
is always present, at every concrescence, in every present occasion. While for British emergentists novel properties rarely
appear, under certain conditions, and are usually explained
by a top-down effect of the global system on its components,
Whitehead's vision appears much more continuous, progressive, through the accumulation of small Darwinian novelties,
and without necessarily calling for feedback loops or action of
the whole on the parts. Finally, the British emergentists seem to
be perfectly compatible with a materialist world view, combin-
ing only the use of the laws of physics with the recognition of
a beyond-the-known laws, initiated by the effects of feedback
or perhaps chance. This is true even if, in Alexander, the evolution of the world is finalised by a deity principle. In contrast,

Whitehead's novelty is introduced through the rupture of temporal continuity fundamentally at work in the process, which consists in the successive replacement of one concrete reality by another, through the sudden and indefinitely repeated crystallisation. It will be seen later that in these crystallisations (or concrescences in Whitehead's language) not only a causal determinism is exercised but also choices, a perhaps infinitesimal but in the long run powerful creative freedom. Thus, Whitehead seems to imply a transcendence, a metaphysics including final laws such as that of the principle of harmony, which Whitehead finally agrees to call 'God'. Here, as on other occasions, if one can understand Whitehead's borrowings from the British pioneers, one must recognise that in this matter, as in others, Whitehead transformed, adapted, and personalised them.

WHITEHEAD AND BERGSON

During his years in London, Whitehead also had the opportunity to be influenced by Bergson. In fact there are many similarities between the French philosopher's thought and Whitehead's vision of the world and about time. The two thinkers showed great esteem for each other. It seems strange that they did not collaborate more. Yet we know that the Whiteheads received Bergson in London on at least one occasion. It may be thought that a dose of mutual intimidation kept them at a respectful distance, Whitehead feeling still a novice in philosophy, and Bergson too alien to developments in the fields of contemporary science, physics, and mathematics.

In terms of similarities, it should be noted first of all that Bergson, like Whitehead, admits a complex and granular status for the notion of time, even if this last particularity seems to be affirmed cautiously and in a veiled manner. One reads for example, in *Matter and Memory*: 'Homogeneous space and homogeneous time are then neither properties of things nor essential conditions of our faculty of knowing them: they express in an abstract form, the double work of solidification and of division which we effect on the moving continuity of the real in order to obtain there a fulcrum for our action, in order to fix within it starting-points for our operation, in short, to introduce into it real changes. They are the diagrammatic design of our eventual action upon matter'.[20] According to Maria Teresa Teixeira, a professor at the University of Lisbon, who has devoted most of her life to a comparative study of Bergson and Whitehead, 'Bergson denies the mathematical continuity of time; time is not infinitely divisible and mathematical continuity is only a discontinuity of points that repeat themselves indefinitely. The duration of the external world, of the physical world, like our own duration, also has a certain thickness, even if it is absolutely invaluable in human terms'.[21]

Whitehead himself comments on Bergson's position on the granularity of time in *Science and the Modern World*:

> Bergson has protested against [the idea of precise location], so far as it concerns time and so far as it is taken to be the fundamental fact of concrete nature. He calls it a distortion of nature due to the intellectual 'spatialisation' of things. I agree with Bergson in his protest: but I do not agree that such distortion is a vice necessary to the intellectual apprehension of nature. I shall in subsequent lectures endeavour to show that this spatialisation is the expression of more concrete

---

20 H. Bergson, *Matière et Mémoire* (Paris: Félix Alcan, 1896), *Matter and Memory*, trans. N. Margaret and W. S. Palmer (London: George Allen and Unwin, 1911), 280.

21 M. T. Teixeira, *Consciência e Acção, Bergson e as neurociências* (Lisbon: Center for Philosophy, University of Lisbon, 2012), 194.

facts under the guise of very abstract logical constructions. There is an error; but it is merely the accidental error of mistaking the abstract for the concrete. It is an example of what I will call the 'Fallacy of Misplaced Concreteness.'[22]

Another similarity is revealed in the treatment of the past. As is well known, in Bergson's work, the past is never completely vanished: it is incorporated into and enriches the present. For Whitehead, as we shall see, the present entities, born of the concrescences that mark the concrete present, weave links with the present entities vanished from the past, which explains both the causality and the permanence of perennial objects. Thus, according to Mrs. Teixeira, it seems 'possible, and even quite probable, that Whitehead developed the idea of an immortal past, which is founded in the actual entities that come and perish, under the direct or indirect influence of Bergsonian philosophy. [...] What Gilles Deleuze has designated, speaking of Bergson's work, as "the contemporaneity of the past and the present".'[23]

Yet another similarity: the refusal to completely dissociate mind and matter, and the extension of the reign of consciousness beyond the human species, or even the reign of the living. Both thinkers believe that everything has a mental aspect, capable of an element of subjectivity. But neither of them wants to be accused of panpsychism, in the sense that an electron would be endowed with the capacity to determine itself. Both admit a continuity between the atom of matter and the spirit of the living, in the sense that even in material bodies, there

---

22 A. N. Whitehead, *Science and the Modern World* (New York: Macmillan, 1925), 52.

23 M. T. Teixeira, *Ser, Devir e Perecer. A criatividade na Filosofia de Whitehead* (Lisbon: Center for Philosophy, University of Lisbon, 2011), 80.

is a possibility of escaping from strict determinism (for how else to explain, according to them, the appearance of novelty?). However, there are differences between the thinking of the two authors. For Whitehead, the mental pole of every actual occasion means that there is in everything an *appetence*, a kind of desire (he uses this word) to conform to the ideal of harmony which is the supreme law of *process*. For Bergson, there is still a difference in nature and not only in degree between the material world and the reign of the living. The latter is subject to a specific law, the famous '*élan vital*', from which the mineral world escapes. The vital impetus, it should be remembered, is the propensity of life itself to create and innovate, particularly with regard to the conservation of the species and the variations that are at the root of evolution.[24]

Consequently, Bergson prefers to reserve the word 'consciousness' for living beings, even though this consciousness can be dormant as in the most primitive animals, bacteria for example. This consciousness admits all kinds of degrees. This gradation is confirmed when we rise in the hierarchy of living beings and, also, when we consider the human consciousness itself.[25]

In Bergson's *Durée et Simultanéité*, published in 1922, we find a laudatory appreciation of Whitehead's *The Concept of Nature*, published two years earlier: 'only recently, in an admirable book, a philosopher-mathematician affirmed the need to admit of an "advance of Nature" and linked this conception with ours'. An added note indicates that *The Concept of Nature* 'is certainly one of the most profound that has been written on the philosophy of nature'. Conversely, in *The Concept of Nature*,

---

24 H. Bergson, *L'évolution créatrice*, trans. A. Mitchell (London: Macmillan, 1907), 89 and following.
25 See for instance Teixeira, *Consciência e Acção*, 18.

Whitehead writes: 'I definitely refrain at this stage from using the word "time", since the measurable time of science and of civilised life generally merely exhibits some aspects of the more fundamental fact of the passage of nature. I believe that in this doctrine I am in full agreement with Bergson, although he uses time for the fundamental fact which I call passage of nature.'[26] However, it should not be forgotten that Whitehead's still influential friend, 'Bernie' Russell, was fiercely opposed to Bergson and found him to be null and void and that Whitehead himself vigorously rejected Bergson's discourse on the 'spatialization of time'. Lowe relates that, when asked at the end of his life, Whitehead replied that he had indeed read Bergson but that he had never really cared about his work.

AFTER THE WAR

On 30 October 1918, Whitehead was elected Dean of the Faculty of Science at the University of London, a position he held for four years, and concurrently served as Chairman of the Academic Council of the University Senate. Figure 3.1 shows his physical appearance, beautiful and inspiring, at that time. As a sign of the times, in 1922 the university created the Department of History and Methods of Science at University College, at the head of which Whitehead held the position of Professor of Logic and Scientific Method (later in 1946, the department was renamed 'History and Philosophy of Science'). In 1923, upon Forsyth's retirement, Whitehead replaced him as head of the Department of Mathematics at Imperial College. It was a promotion, but Whitehead did not like the attention he was supposed to give to all the administrative details of his accumulated duties and regretted not being able to devote as

---

26 A. N. Whitehead, *The Concept of Nature* (Cambridge: Cambridge University Press, 1920), 93.

Figure 3.1. *Whitehead, undated portrait, likely during his stay in London, circa 1920.*[27]

much time as he would have liked to reflection and research. All the more so as at this time he was also asked by the English

---

27 Lowe, *Alfred North Whitehead*, Vol. II, 181.

authorities to reflect on the place of mathematics and natural sciences in teaching.[28]

But let us return to Whitehead's concerns about a philosophy of nature and time that would be based solely on perceptual data, in his post-war London period. In Victor Lowe's opinion, thinking about Time is what stimulated him most, and during this period of his life, Einstein's theory had an enormous influence on him.[29] On this subject, we already know that Whitehead gave four lectures on Relativity to the Aristotelian Society between 1915 and 1922; he also helped his compatriot Richard B. Haldane (Lord Haldane) to understand Einstein's theory and to publish a book about it in 1921 entitled *The Reign of Relativity*. He himself published in 1919 his book *An Enquiry about the Principles of Natural Knowledge*, which contains, apart from a systematic exposition of the theory of extensive abstractions, numerous references to the theory of Relativity, and already some criticisms of Einstein's theory; he develops these in his new publication of 1922 (*The Principle of Relativity with Applications to Physical Science*), accompanied by a reflection undoubtedly intended to prevent any accusation of immodesty: 'the worst homage we can pay to genius is to accept uncritically formulation of truths which we owe to it'.[30]

Thus, it is reasonable to assume that throughout his London years, Whitehead devoted much time and energy to exploring the nature of Time, opened by his letter to Russell in September

---

28 Lowe, *Alfred North Whitehead*, Vol. II, 71–78, 84, 102. When Whitehead published *The Aims of Education and Other Essays* in 1929, he continued his reflections in the same vein, protesting against the habit of giving students stupid problems, pleading to promote personal reflection rather than automatic mechanisms, and to relativise the importance of merit rankings.

29 Lowe, *Alfred North Whitehead*, Vol. II, 123.

30 A. N. Whitehead, *The Principle of Relativity with Applications to Physical Science* (Cambridge: Cambridge University Press, 1922), conclusion of the first part.

1911. It seems to me that we can distinguish four stages, or four kinds of considerations. First, the distinction between abstract time and process, this perpetual and inherent unrest of nature itself; second, the highlighting of the singularity of the act of concretisation of reality, which is also an act of permanent but discrete creation (in the mathematical sense of this term), through the succession of concrescences; it naturally follows an 'atomisation' of time, that only the bridges thrown between the actual occasions of the past and the concrescences in the process of realisation save from pure contingency; and finally the discovery by the human mind of the extensive space–time continuum, the only way to apprehend the true ultimate constituents of nature that are the particle-events and to understand the irreducible interweaving of the too human concepts of space and time.

As an illustration of the first of these four considerations, let us note the following passage in *The Concept of Nature*:

> Time is known to me as an abstraction from the passage of events. The fundamental fact which renders this abstraction possible is the passing of nature, its development, its creative advance, and combined with this fact is another characteristic of nature, namely the extensive relation between events. These two facts, namely the passage of events and the extension of events over each other, are in my opinion the qualities from which time and space originate as abstractions.[31]

The third explains how Whitehead 'secured the notion of causation from Hume's skepticism', to use Victor Lowe's words, and restores the status of the relative permanence of objects: 'events are here and now; they happen once and do not repeat

---

31　Whitehead, *The Concept of Nature*, 71.

themselves. Objects, on the other hand, are the *recognita* discriminated in any complex of events.'[32]

DEPARTURE FOR AMERICA

After the war, the Whitehead couple no longer travelled to Europe, contrary to their pre-war habits. In 1922, they went to America for the first time, to participate in the celebration of the retirement of Charlotte Angas Scott, Professor of Mathematics at Bryn Mawr College, near Philadelphia. Probably, if Whitehead accepted the invitation to deliver the celebratory lecture there, it was not only because of the scientific merits of Charlotte Angas, but also in memory of the fact that she had been the first woman to compete in the tripos exam at Cambridge, had obtained a score that would have designated her as eighth wrangler, but had not obtained it officially because at that time Cambridge did not award diplomas to women candidates (Whitehead has always been an active promoter of higher education for women). They came back enchanted and conquered by this trip to America. Two years later, a group of American friends, learning that Whitehead would soon be to be in position of retirement from the University of London, decided that they should try to attract him to Harvard and have him appointed as a Professor in the Department of Philosophy at that University. The proposal was made to the President of Harvard, Lawrence Lowell, with the support of historian Henry Osborn Taylor, who even offered to pay the salary of the chair for a few years.[33] The President of Harvard wrote to Whitehead to propose it. When the letter arrived in London, Alfred had

---

32 Lowe, *Alfred North Whitehead*, Vol. II, 130, 116.
33 W. E. Hocking, *Alfred North Whitehead, Essays on His Philosophy* (Lanham: University Press of America, 1989), 10–11.

Evelyn read it to him, and Evelyn asked her husband, 'What do you think of it?' To her surprise, she heard him say, 'I would rather do that than anything else in the world'. In fact, Taylor paid Harvard the equivalent of Whitehead's salary until Whitehead retired in 1937. But the Whiteheads did not know this until the death of their protector in 1941.[34]

---

34 Lowe, *Alfred North Whitehead*, Vol. II, 133–35.

CHAPTER 4

The Instant Does Not Exist: Rebuttal of Descartes' and Newton's Time

> It needs very little reflection to convince us that a point in time is no direct deliverance of experience. We live in durations, and not in points.
>
> A. N. Whitehead, *The Aims of Education and Other Essays*, 1929

In his 1905 article, "On Mathematical Concepts of the Material World", Whitehead proposed new basic concepts to describe the world around us. In its first part, he considers the classical physics of Descartes and Newton. He explains that classical physics has imposed the trilogy: *points of space, particles of matter*, and *instants of time* as basic ingredients. For classical science, the point of space is understood as a necessary concept for the development of geometry, the ultimate reality for decomposing space. Matter is naturally the substance of the world, and the concept of the material point is necessary to be able to precisely locate this matter in space. But matter can be at rest or in motion in this space. Therefore, to describe motion, we must use the common concept of time. And finally, to describe nature, it is necessary to be able to precisely locate each material point at each instant, whose concept is similar to that of a point in space but relative to time.

Taking into account the difficulties encountered in his joint work with Bertrand Russell on the notion of point as the basis of geometry (see Chapter 2), Whitehead proposes in this seminal article to follow rather the inspiration of Leibniz and his 'relational' conception of space. This leads him to propose as ultimate elements of reality not points and particles, but lines (lines of force) in which geometrical points are only elusive abstractions.

In this vision, the world scene still unfolds continuously in this classical time. Whitehead even asserts, at the beginning of the article, that 'every concept of the material world must include the idea of time. Time must be composed of *Instants*'. This 'must' is emphasised by Whitehead's reference in this publication to Bertrand Russell's *Review of Kant's Cosmogony*, published in 1901. And Whitehead goes on to say: 'Thus *Instants of time* will be found to be included amongst the ultimate existents of every concept [which I will examine in this article]'. He then reviews five classes of possible systems, from the classical Newtonian system to a specifically Whiteheadian (relational) system, but in which ultimate realities always include instants of time.

Despite the very 'classical' nature of this article with regard to the metaphysics of time, it marks Whitehead's first publication in the field of metaphysics and philosophy of science. Whitehead is 46 years old. His interest in philosophy and metaphysics has gradually grown, but has not yet reached its fullness. He is perhaps still partly trapped by the influence of Bertrand Russell, his former student, and at that time his closest friend and collaborator in the writing of *Principia Mathematica*. Russell, in those years, was indeed considered a specialist of Leibniz, while remaining attached to the classical vision of time. Moreover, in 1905 and in the years that followed, Whitehead was still fully occupied with his administrative and teaching duties at Trinity College, as well as with the writing (with Bertrand Russell) of the *Principia*.

Three years later, Hermann Minkowski opened his lecture in Cologne on 21 September 1908 with these words: 'Gentlemen! The concepts about time and space, which I would like to develop before you today, have grown on experimental physical grounds [...] Henceforth, space for itself, and time for itself shall completely reduce to a mere shadow, and only some sort of union of the two shall preserve independence'. This is the public birth of four-dimensional space-time. In this mathematical continuum, time is treated on an equal footing with space, except for two details: time is made homogeneous with space by its multiplication with the constant c, the speed of light, and by a sign -, which has its importance. But the articles published by Minkowski in 1907–1909 did not attract Whitehead's attention. He later confided to his biographer Victor Lowe that he did not pay attention to these articles until about ten years later.

In 1910, the Forsyth affair broke out in Cambridge, and the Whitehead couple decided to move to London. Their installation completed, the course of life resumed. In September 1911, Whitehead was correcting the proofs of the second volume of the *Principia* and working on the theme of volume IV, geometry. Very late in the evening of 2 September, he suddenly had an inspiration, an illumination, which he delivered the next day in a letter to Bertrand Russell:

> Last night, when I should have finished [the revisions of the second volume], the idea suddenly flashed on me that time

could be treated in exactly the same way I have now got space (which is a picture of beauty, by the bye). So till the small hours of the morning, I was employed in making notes of the various ramifications. The result is a relational theory of time, exactly on four legs with that of space. As far as I can see, it gets over all the old difficulties, *and above all abolishes the instant in time.*[1]

Whitehead is now 50 years old, the age of maturity. He will finally be able to devote more time to the elaboration of his system of the world, which he wants to build on the sole basis of primary sensitive data, and not on abstractions. But he does not leave his letter to Russell without adding a paragraph to confess that this history of the nature of time has, in fact, always bothered him, because he has always suspected that time was a much more complex property than the common notion, 'phagocytised' so to speak by Galileo to place it at the centre of his vision of World Mechanics, to the detriment of space,[2] and axiomatised by Descartes and especially Newton. 'But I have had to conceal my dislike from lack of hope. But I have got my knife into it at last' (letter from Alfred North Whitehead to Bertrand Russell, 3 September 1911).

At that time, it should be remembered, Whitehead was fighting for a better job than the newly conceded position of lecturer in Applied Mathematics and Mechanics. In July 1912, he obtained the position of Assistant Professor (Reader) in Geometry, still at University College. And two years later, he moved from there to the Imperial College of Science and Technology, where he finally obtained the position of Professor

---

1   V. Lowe, *Alfred North Whitehead, the Man and His Work, Volume I: 1861–1910* (Baltimore: The John Hopkins University Press, 1985), 299. Italics added by R. L.

2   See R. Lestienne, *The Children of Time* (Urbana: University of Illinois Press, 1995), chap. 3.

of Applied Mathematics, thanks in particular to the support of Andrew Forsyth, who in the meantime had been appointed Chief Professor of Mathematics in this newly created institution. During this time, he continued his reflections in logic, mathematics, and metaphysics, gave lectures in mathematics and education (a recurring theme in Whitehead's reflections and publications), but still refrained from publishing anything in philosophy of science.

In April 1914, however, he took part in the *Philosophie des Mathématiques* congress in Paris, where he read his contribution on '*la théorie relationnelle de l'espace*', which we have already mentioned in the previous chapter. This article confirms his inclination for a Leibnizian approach to space, far from the absolute Newtonian conception, but does not discuss the notion of time and says nothing about dynamics.

Four months later, war broke out, and his two sons were successively engaged in battle. A month later, he takes up his post at Imperial College. In March 1918, the family was devastated by Eric's death. Six months later, Whitehead was elected Dean of the Faculty of Science at the University of London. Wartime worries, administrative and teaching duties, and perhaps most of all the grief that befell the family, delayed the writing of books on the philosophy of science, which he nevertheless thought about more and more.

His first return to this field was marked by the book *An Enquiry Concerning the Principles of Natural Knowledge*, published in 1919, followed a year later by the publication of *The Concept of Nature*.

An Enquiry Concerning the Principles of Natural Knowledge opens with a dedicatory page to the memory of his son Eric, a testimony to the shock of this disappearance for the Whitehead family. In this book, Whitehead finally addresses the question of the existence or non-existence of instants. His first chapter is already a severe criticism of the foundations of classical science, based, as he reminds us, on three essential elements, the point, the instant, and the particles of matter, ordered within the framework of an absolute space (composed of points) and a universal time (composed of instants).

With regard to time, he writes:

> We must therefore in the ultimate fact, beyond which science ceases to analyse, include [among the essential physical quantities] the notion of a state of change. But a state of change at a durationless instant is a very difficult conception. It is impossible to define velocity without some reference to the past and the future... This conclusion is destructive of the fundamental assumption that the ultimate facts for science are to be found at durationless instants of time.

Indeed, how can we fail to follow him in this common sense deduction: in this world of incessant change, of permanent flow, of obvious dynamisms, how can we believe that a scientific representation of reality can be based on static snapshots? This is why our mathematician–philosopher is continuing on his path:

> Accordingly it is admitted that the ultimate fact for observational knowledge is perception through a duration; namely, that the content of a specious present, and not that of a durationless instant, is an ultimate datum for science [...] Our perception of time is as a duration, and these instants have only been introduced by reason of a supposed necessity of thought. In fact absolute time is just as much a metaphysical monstrosity as absolute space.[3]

---

3   A. N. Whitehead, *An Enquiry Concerning the Principles of Natural Knowledge* (Cambridge: Cambridge University Press, 1919), 2–8.

In this passage, a word must hold our attention, because it points to one of the probable sources of Whitehead's thinking, at least one of the arguments evoked for its support: the *specious present*, that is to say, susceptible of creating an illusion by its appearance of truth, of adequacy to reality. For philosophers, this expression immediately refers to the work of William James, the great Harvard psychologist and philosopher, who in 1890 published his treatise *The Principles of Psychology*.

The chapter he devoted to the perception of time contrasts the concrete experience of the present with the abstract present, the mathematical instant without duration separating the past and the future. Apparently quoting one of his acquaintances, James writes: 'Let it be named the *specious present*, and let the past, that is given as being the past, be known as the *obvious past*. All the notes of a bar of a song seem to the listener to be contained in the present. All the changes of place of a meteor seem to the beholder to be contained in the present'. And he concludes: 'In short, the present that we practically know is not thin like the blade of a knife, but is shaped like a horse's saddle, with a certain width of its own on which we are sitting, and from which we look in both direction of time'.[4]

In Chapter XV of his book, James expands on the reasons for introducing this thick present. He cites the observations of several psychologists of his time, Wundt, Fechner, and others. Some sought the maximum span of time such that the events

---

4   James therefore does not claim authorship of the specious present expression, although today it is indissolubly attached to his name. See H. Andersen, "The Development of the 'Specious Present' and James' Views on Temporal Experience" 2014, http://philsci-archive.pitt.edu/10721/. For his part, one might think that Whitehead became familiar with the thought of William James through his attendance in London at the Aristotelian Society, and especially through his exchanges with Charles Broad, a notorious disciple of James.

that compose it (e.g. a succession of musical notes) are felt to constitute a distinct whole. Their conclusion varies from a few seconds to about a dozen. Others, on the contrary, have looked for the minimum time span for two events to be perceived as distinct. This interval obviously depends on the sense organ concerned: Exner finds that two claps of sparks can be perceived as successive if the interval between them exceeds about 2 milliseconds, but that two flashes of light falling on the same region of the retina are only distinguished if they are separated by more than 44 milliseconds. Remember that modern cinema uses cameras filming at a rate of 24 frames per second, which leaves 42 milliseconds per frame. Others, finally, have investigated what is the smallest difference between time intervals that can be recognised, and what is the time interval for which this difference in elapsed time is best appreciated, in other words, the period of time that is most accurately assessed. Most of the researchers cited by James agree on a duration of the order of 750 milliseconds.

> We saw a while ago, summarizes James, that our maximum distinct *intuition* of duration hardly covers more than a dozen seconds ... we must suppose that *this amount of duration is pictured fairly steadily in each passing of consciousness* by virtue of some fairly constant feature in the brain-process to which the consciousness is tied. [...] The duration thus steadily perceived is hardly more than the 'specious present', as it was called it a few pages back. [...] *Duration and events together form our intuition of the specious present with its content.* Why such an intuition should result from such a combination of brain-processes? I do not pretend to say. All I aim at is to state the most *elemental* form of the psycho-physical conjunction.[5]

---

5 W. James, *The Principles of Psychology* (New York: Henry Holt & Comp., 1890), chap. XV.

Psychophysiology was certainly not sufficiently developed at that time to specify and understand the physiological mechanisms concerning the appreciation of durations and the content of the specious present. Nowadays, research in neuropsychology and neurophysiology has confirmed and largely clarified James' reflections. Much more than being a machine for taking photographic snapshots of external reality, our brain today appears as a machine for reviewing sensory data to ensure their coherence and analyse their probable causes, in order to determine the actions to be taken. Such an evaluation necessarily takes time. Verifying the coherence of the sensations received, the brain is able to obliterate details that seem surreal, or even add others that seem missing, in order to accredit the causes of the scene experienced. For example, contemporary researchers have shown that a fake gorilla crossing the stage of a theatre where a captivating scene between several characters is taking place is generally not seen by the spectators, while a person used to watching passers-by in the street from his window will not see that his neighbour usually wearing a hat has gone out bare-headed today. Moreover, Benjamin Libet's famous experiments, clarified and more carefully planned by other researchers, have shown that the nervous system prepares the execution of a voluntary movement several hundred milliseconds before the subject becomes aware of his or her willingness to perform it. In the same vein, other researchers have shown that the order of temporal succession of sensory stimuli (auditory, visual, tactile) which follow one another rapidly depends largely on their context, sometimes seeming to the subject to anticipate their actual occurrence by several hundred milliseconds. These mechanisms, for correcting the sensations received, explain, for example, why in a concert hall, when a musician strikes his cymbals, we have the sensation of simultaneity between the act seen and the sound emitted; whereas, as we know, sound only travels at a speed of 330 m/s: the image therefore precedes the sound received by several tens, or even a hundred milliseconds, depending on the distance between the spectator and the orchestra.

All these observations converge today and make it possible to affirm that what we call our visual present is not a photographic snapshot of the 'true' reality but a composite painting that can be revised (a little like the famous painting of Napoleon's coronation!) and constructed within a window of time of the order of half a second.[6]

A year after the publication of his book *An Enquiry Concerning the Principles of Natural Knowledge*, Whitehead published *The Concept of Nature*. This new book contains an entire chapter devoted to Time, the title of which is simply 'Time'. Essentially, this chapter is Whitehead's written transposition of a lecture he gave in the fall of 1919 at Trinity College, Cambridge.

The specious present is again present on every page of this book, with deep resonances with James' reflections. Whitehead writes:

> Instantaneousness is the concept of all nature at an instant, where an instant is conceived as deprived of all temporal extension. For example we conceive of the distribution of matter in space at an instant. This is a very useful concept in science especially in applied mathematics; but it is a very complex idea so far as concerns its connexions with the immediate facts of sense-awareness. There is no such thing as nature at an instant posited by sense-awareness.

---

6   See for example, R. Lestienne, "The Duration of the Present," in *Time, Perspectives at the Millennium,* ed. M. P. Soulsby and J. T. Fraser (Westport: Bergin & Garvey, 2001), chap. X; R. Lestienne, *Le Cerveau Cognitif* (Paris: CNRS Editions, 2016).

What sense-awareness delivers over for knowledge is nature through a period.

And, a little further on, we find this passage with particularly 'Jamesian' accents:

> The character of a moment and the ideal of exactness which it enshrines do not in any way weaken the position that the ultimate terminus of awareness is a duration with temporal thickness. This immediate duration is not clearly marked out for our apprehension. Its earlier boundary is blurred by a fading into memory, and its later boundary is blurred by an emergence from anticipation.[7]

Another aspect of this lecture should attract our attention, as it prepares the separation of Whitehead's philosophy of nature from Albert Einstein's, that of the theory of Relativity. Whitehead indeed devotes long developments to defining the relation of simultaneity. Far from rejecting this concept, Whitehead makes it a general fact that is binding on any observer. It is 'all nature now present as disclosed in [our] sense-awareness... These are the events whose characters together with those of the discerned events comprise all nature present for discernment. [...] Simultaneity must not be conceived as an irrelevant mental concept imposed upon nature. Our sense-awareness posits for immediate discernment a certain whole, here called a "duration" [...] A duration is a concrete slab of nature limited by simultaneity which is an essential factor disclosed in sense-awareness'. He proposes to call this a 'moment'. Naturally, Whitehead agrees that a distinction must be made between simultaneity thus defined and instantaneousness. We shall

---

7   A. N. Whitehead, *The Concept of Nature* (Cambridge: Cambridge University Press, 1920), 24, 27.

come back to these points by evoking Whitehead's development of a theory of relativity distinct from that of Einstein.

***

But let us come to the other concrete pillar of his defence of the thick present and the unreality of the conceptual instant. This was brought to him, in the second part of the 1920s, by the development of quantum mechanics, then often called 'wave mechanics'.

In 1911, quantum mechanics was still in limbo. It was not until 1925 that the main principles of the new science were laid down by Werner Heisenberg, Erwin Schrödinger and the physicists who surrounded Niels Bohr in Copenhagen. It is remarkable that Whitehead took up the subject as early as that year 1925, in his book *Science and the Modern World*, when the foundations of quantum mechanics were still under discussion amongst the main protagonists. At that date, Heisenberg had not yet officially put forwards his vision of measurement, which would be at the origin of the postulate of 'collapse of the wave function', which broke with the continuity in the evolution of quantum systems and introduced indeterminism into physics. This shows that, contrary to the generally accepted view, Whitehead was very attentive to developments in this new discipline and its implications for the philosophy of nature.

Let us recall what this is all about. Let us consider a microscopic system (e.g. an electron) whose property P is to be measured. Wave mechanics posits that the physical state of this system is represented by a wave function, ψ, which is generally the sum of several 'eigenfunctions' ψ_k each representing the state of the system for which P has the value 'k': $\psi = \Sigma c_k \psi_k$ (this is the Principle of superposition, which has no equivalent in classical mechanics, because in the latter, each of the measurable properties of each object has a defined value). Under these conditions, quantum mechanics says that the probability that the measurement gives the result k is given by the coefficient c_k of the previous development, squared: $\Pr(k) = c_k c_k^*$. Moreover, after a measurement which gave the value 'k' for the property P, the wave function representing the system has suddenly changed, it is now the wave function ψ_k: there has been a collapse of the wave function. For the founders of wave mechanics, this collapse is conceived as purely instantaneous.[8]

In *Science and the Modern World*, Whitehead writes on this subject:

The point is that one of the most hopeful lines of explanation is to assume that an electron does not continuously traverse its path in space. The alternative notion as to its mode of existence is that it appears at a series of discrete positions in space which it occupies for successive durations of time. It is as though an automobile, moving at the average rate of thirty miles an hour along a road, did not traverse the road continuously; but appeared successively at the successive milestones, remaining for two minutes at each milestone. [...] The discontinuities introduced by the quantum theory require revision of physical concepts in order to meet them. In particular, it has been pointed out that some theory of discontinuous existence is required. What is asked from such

---

8   Today, it is thought that the collapse of the wave function, which characterises the passage from the field of validity of quantum mechanics to classical mechanics, is linked to the interactions of the microscopic system with the measuring apparatus and the environment, and is not really instantaneous but is all the more rapid as these interactions concern more 'massive', more complex systems (measuring apparatus, environment).

a theory is that an orbit of an electron can be regarded as a series of detached positions, and not as a continuous line.[9]

He goes on to explain that a complete period of the wave associated with an object event defines the duration required for the complete manifestation of its structure. Thus, the primary event is realised in an atomic way in a succession of durations, each duration having to be measured from one maximum of the wave to the other. Thus, as far as this event is seen as a persistent entity must be taken into account, its persistence must be assigned to these durations taken successively. If it is considered as a thing, its trajectory must be mathematically represented by a series of detached points. So the movement of a primary event is discontinuous in space and time.

The quantum conception of motion is illustrated in a particularly suggestive way in the pictures of traces of elementary particles such as electrons in a cloud chamber. This device, which was implemented in England in the same years by Charles Wilson, and for which he was awarded the Nobel Prize[10] in 1927, makes it possible to visualise the trajectory of quantum particles in motion. More precisely, a cloud chamber is a glass enclosure that contains the vapour of a chemical substance at a partial pressure close to its condensation value. The particles that pass through it, for example, from cosmic rays, ionise this gas and cause tiny fog droplets to form along their path. By placing the Wilson chamber in the air gap of an electromagnet, one can observe the curvature of the trajectories under the influence of the magnetic field of the electromagnet

---

9   A. N. Whitehead, *Science and the Modern World* (New York: Macmillan, 1925), 36, 137.

10   In particular, Wilson's chamber enabled the first observation of a positron (positively charged electron) by Carl Anderson in 1932.

and deduce the velocity of the particles in question. Now, what do the trajectories of these particles look like in a cloud chamber? Precisely, as a discontinuous series of droplets, just like Whitehead describes in his book the trajectory of a 'quantum' automobile.

Figure 4.1 is an example of a Wilson's chamber picture. It shows in the centre the collision of a positron (positive electron, arriving from below) with an atomic electron, resulting in

Figure 4.1. *Photograph taken in a cloud chamber, allowing to visualise the trajectory of elementary particles in a magnetic field. These trajectories are not continuous but are formed by a string of separate droplets.*

the deflection of the positron and the ejection of the electron, recognisable by the opposite curvature of their trajectories in the magnetic field. One can clearly distinguish in this picture the beads of successive droplets that reveal the trajectory of the different particles involved.

IF THE INSTANT DOESN'T EXIST, THEN THE TIME OF DESCARTES AND NEWTON DOESN'T REPRESENT REAL TIME, IT'S AN ABSTRACTION, AND A MUTILATION.

For Whitehead, the time of Descartes and Newton is inseparable from the conception of matter as ultimate reality. Indeed, time is only introduced into physical theory to account for the movement of matter particles. But for Whitehead, matter is not the ultimate reality. A particle of matter is not, strictly speaking, a substance, that is, what needs only itself to exist, to paraphrase Descartes. For its persistence through these bubbles of existence that we have been talking about is relative; it depends on its relations with the other 'substances' that, closely or remotely, influence it. Everything is a matter of relationships in Whitehead; his vision of nature, geometry, space, and time is decidedly Leibnizian.

Addressing the question of time in *Science and the Modern World*, Whitehead begins by paying tribute to Descartes:

In his distinction between time and duration, and in the way he bases time on motion, and in the close relationship he attributes to matter and extension, Descartes anticipates, as strongly as was possible in his time, the modern notions suggested by the theory of Relativity, or in some aspects of Bergson's doctrine on the generation of things.[11]

11 Whitehead, *Science and the Modern World*, 145.

Before Descartes, Whitehead continued, the worlds of matter and spirit shared the experience of general flow; 'but time, as measured, is assigned by Descartes to the cogitations of the observer's mind. There is obviously one fatal weakness to this scheme. [...] The subject–object conformation of experience in its entirety lies within the mind as one of its private passions'.[12] Indeed, Descartes devoted, scattered throughout his work, long reflections on time and its relationship with the mind. He firmly recognises that things are born in time and that time is therefore an existential attribute of things, and he goes even further by affirming the permanent re-creation of things and beings by God throughout this duration. But, in accordance with the dichotomy between extension and thought or between things and spirit, Descartes reduced time as a passage exclusively for spirits: for him, time is linked to memory, it is the tool that allows us to measure the duration of things. In contrast, Whitehead wants to link the passage of nature, birth, death, and the re-actualisation of events to nature itself to the *process*. This is why he concludes his discourse on time in *Science and the Modern World* by stating that the context of the discoveries that have emerged in previous years concerning matter, space, time, energy has become so complex that it has destroyed the security that surrounded the old catalogue of orthodox principles (matter, points of space, moments of time) and that it is therefore necessary to reorganise all this.

As for Newton's space and time, let's remember that they just form the framework for the unfolding of the world scene; they have nothing to do with the events taking place there. But why, if the instant does not exist, are Newton's laws, which involve first (speed) and second (acceleration) derivatives with respect to time, so effective?

---

12 Ibid., 146.

The effectiveness of abstraction should not be denied. Newton's laws are efficient and it is permissible to construct, as Newton and Leibniz did, first and second derivatives of any continuous function. But the t-parameter of Newtonian mechanics only represents a distant abstraction from real time, that is, from the universal flow in which we are immersed. It is a kind of average obtained by jumps that erase the atomic accidents of time, thanks to a statistical law of causality whose foundations according to Whitehead will be analysed in Chapter 5.

Convinced of the pitfalls of contemporary language, accentuated by the developments of classical science, Whitehead decided to break with our language habits. In *The Concept of Nature*, he announces: 'I definitely refrain at this stage from using the word "time", since the measurable time of science and of civilised life generally merely exhibits some aspects of the more fundamental fact of the passage of nature'. And finally, in *Science and the Modern World*, he introduces the term that will henceforth have his preference: *process*. 'In the analogy with Spinoza, his one substance [of which Spinoza is talking about] is for me the one underlying activity of realisation individualising itself in an interlocked plurality of modes. [...] *Thus, concrete fact is process. [...] Thus, nature is a structure of evolving processes. The reality is the process*'.[13]

---

13 Ibid., 87–90. Final italics added by R. L.

CHAPTER 5

The Successive Crystallisations or 'Concrescences' of Reality

The first and most important consequence of the disappearance of the instant and the 'atomisation of time' into elementary bricks is that change in nature is now understood as not being continuous. It is a succession of fleeting tableaux, a string of bubbles; with each bubble of evolution, there is room for the appearance of novelty and thus, in the course apparently dictated by the determinism of events, the possibility for bifurcations to appear.

So how does Whitehead see the world around him? Let us imagine him in the small second home he had built in 1929 about 20 kilometres south of his home and Harvard University. His window is open to the large meadows, where a few horses are quietly grazing. To the right, a path leads up to the nearby Blue Hills, which are very wooded and draw a horizon of winding curves and peaceful relief. As he looks up from the manuscript he is working on, a doe quietly crosses the meadow, occasionally glimpsing the small house. Whitehead sees this bucolic scene as we would see it, enjoying its fullness as we would admire its bluish, green, and purple tones. But he interprets his vision differently from most of us.

Whitehead explained to Lucien Price that although our judgments today are closely conditioned by common notions of space and time, in truth the present reality is outside of time and space: it is the process itself that is the present reality.[1] And in *The Aims of Education* he adds this thought which has a Kantian resonance: 'It is not true that we are directly aware of a smooth running world [...] In my view the creation of the world is the first unconscious act of speculative thought; and the first task of a self-conscious philosophy is to explain how it has been done'.[2]

As we have seen, for Whitehead, the course of change, the *process*, is analogous to a vibration. A vibration must be complete in order to exist as a vibration, like a musical note. Each vibration leads to the birth of a new reality. We therefore live in a sequence of crystallisations of reality, or as Whitehead says, of 'concrescenses'. This word appears for the first time in a lecture given at the Sixth International Congress of Philosophy, held at Harvard from 13 to 17 September, 1926. Whitehead gave a paper entitled "Time". In it, he states first of all that 'if time is taken seriously, no concrete entity can change: it can only be replaced'. Thus, concrete entities replace each other indefinitely.

CONCRESCENCES

So far, is the idea so original? Most of us will agree that the only concrete, current realities are those of the present. Let us

---

1   L. Price, *Dialogues of Alfred North Whitehead, As Recorded by Lucien Price* (New York: The New American Library, 1954), 174.

2   A. N. Whitehead, *The Aims of Education and Other Essays* (New York: Macmillan, 1929), 163–64.

think back to the reflections of Saint Augustine in Book XI of his *Confessions*:

> How is it that there are the two times, past and future, when even the past is now no longer and the future is now not yet? But if the present were always present, and did not pass into past time, it obviously would not be time but eternity. If, then, time present—if it be time—comes into existence only because it passes into time past, how can we say that even this is, since the cause of its being is that it will cease to be? Thus, can we not truly say that time *is* only as it tends toward nonbeing?[3]

We find a similar idea in Descartes: 'The truth of this demonstration will clearly appear, provided we consider the nature of time, or the duration of things; for this is of such a kind that its parts are not mutually dependent, and never co-existent; and, accordingly, from the fact that we now are, it does not necessarily follow that we shall be a moment afterwards, unless some cause, viz., that which first produced us, shall, as it were, continually reproduce us, that is, conserve us.'[4] In a word, then, the persistence of a being or a thing is nothing other than a perpetual re-creation by God. Whitehead, as for him, prefers to refer to Locke, whom he considers closer to his own conception, although one can estimate that on this point he did not say anything other than Descartes. In the *Essay on Human Understanding*, Locke indeed writes, 'Duration, and time which is a part of it, is the idea we have of perishing distance, of which

---

3   St. Augustine, *Confessions*, Book XI, chap. XIV (tr. Albert C. Outler, Phildelphia: Westminster Press, 1955).
4   R. Descartes, *Principles of Philosophy*, Book 1, 21 (tr. John Veitch, New York: Barns & Noble, 1644/2009).

no two parts exist together, but follow each other in succession".[5] Whitehead's interpretation of the sensory world did not come out of nowhere. Our generation bathed in movies has no great difficulty in seeing in this art the close analogue of Whitehead's interpretation of reality. Through a rapid succession of still images, cinema knows how to give us the illusion of movement. Of course, this illusion is offered to us thanks to the remanence properties of the retina. In Whitehead's philosophy, we have to assume that the rhythm of succession of these concrescences is much faster still, completely imperceptible. In other words, for him the universe itself is first of all a state of potentiality, which is concretised in successive concrescences that occur, so to speak, out of time. These concrescences make the universe become real and concrete in the sense that we give to this term, that is to say, observable and which can be manipulated by us. Not only the universe as a whole, but every object around us undergoes the same incessant process of concrescences. An electron is a series of microscopic concrescences, a clock or a chair are a series of mesoscopic concrescences, and the whole universe is as much at the cosmological level. And each concrescence adds a parcel of time to the local clock.

Thus, for Whitehead, the scene he has just experienced in the countryside is, in the first sense, a rapid succession of frozen tableaux. A series of sudden crystallisations of a world of fluid virtualities—of potentialities. One could say that they are like flashes, and that our vision is thus chopped, stroboscopic, as when in a dance hall plunged into darkness we switch on the spotlights intermittently and closely, but that would be going too far. For, on the other hand, as he stated so forcefully, reality

---

5   J. Locke, *An Essay Concerning Human Understanding* (Pennsylvania: The Pennsylvania State University, 1690/1999), XV, § 12.

is process, fluency, change, creativity. 'The doctrine is founded upon three metaphysical principles. One principle is that the very essence of real actuality—that is, of the completely real—is the *process*. Thus each actual thing is only to be understood in terms of its becoming and perishing. There is no halt in which the actuality is just its static self, accidently played upon by qualifications derived from the shift of circumstances. The converse is the truth.'[6]

How can these two aspects, that of the process and of the frozen tableaux, be held together?

In *Process and Reality*, Whitehead builds an explanation: 'At this point the group of seventeenth- and eighteenth-century philosophers practically made a discovery, which, although it lies on the surface of their writings, they only are half-realised. The discovery is that there are two kinds of fluency. One is the *concrescence* which, in Locke's language, is 'the real internal constitution of a particular existent'. The other is *the transition* from particular existent to particular existent. This transition, again in Locke's language, is the 'perpetually perishing' which is one aspect of the notion of time; and in another aspect, the transition is the origination of the present in conformity with the 'power' of the past.

The phrase 'the real internal constitution of a particular existent', the description of the human understanding as a process of reflection upon data, the phrase 'perpetually perishing', and the word 'power' together with its elucidation are all to be found in Locke's *Essay*. Yet owing to the limited scope of his investigation, Locke did not generalise or put his scattered ideas together. This implicit notion of the two kinds of flux finds

6 A. N. Whitehead, *Adventures of Ideas* (New York: Free Press, 1933/1967), 274.

further unconscious illustration in Hume. It is all but explicit in Kant, though—as I think— misdescribed. Finally, it is lost in the evolutionary monism of Hegel and of his derivative schools. With all his inconsistencies, Locke is the philosopher to whom it is most useful to recur, when we desire to make explicit the discovery of the two kinds of fluency, required for the description of the fluent world. One kind is the fluency inherent in the constitution of the particular existent. This kind I have called 'concrescence'. The other kind is the fluency whereby the perishing of the process, on the completion of the particular existent, constitutes that existent as an original element in the constitutions of other particular existents elicited by repetitions of process. This kind I have called 'transition'.[7]

INCORPORATING THE THICKNESS OF THE PAST INTO THE PRESENT

Concrescence incorporates things from the past into the reality of the present. 'Our lives are dominated by enduring things', writes Whitehead, 'each experienced as a unity of many occasions bound together by the force of inheritance'. So we need to complete our vision of concrescence, of transition, in a word, of process. We must make room for the immersion of the past in the present, or as Whitehead says, for the power and objective immortality of the past. How can this be done? Let us continue to quote our author in *Adventures of Ideas*: 'It is here that the Aristotelian doctrine of primary substances has done some of its worst harm. For according to this doctrine no individual primary substance can enter into the complex of objects observed in any occasion of experience [...] The individual, real

---

7　A. N. Whitehead, *Process and Reality* (New York: Macmillan, 1929/ New York: Free Press, 1979), 210.

facts of the past lie at the base of our immediate experience in the present. They are the reality from which the occasion springs, the reality from which it derives its source of emotion, from which it inherits its purpose, to which it directs its passion'.[8]

The reader will no doubt find that these words resonate strongly with those of Bergson, who insisted so much on the persistence of the past in the present, and he will be right. Bergson is indeed the one who denied the adequacy of the representation of time as a line of successive points. It was he who wrote in *The Creative Evolution*: 'My mental state, as its advances on the road of time, is continually swelling with the duration which it accumulates: it goes on increasing—rolling upon itself, as a snowball on the snow', and a little further on, 'The past remains bound up with the present'.[9] Whitehead had read Bergson well, knew him well, and even, as we already know, received Bergson at his home in London. But he goes further. Where Bergson reasons according to human consciousness and memory, he extends the solidarity of the past with the present in the very ontology of becoming, in the very mechanism of concrescence, as we will develop in Chapter 6.

A LOOK BACK AT THE CONCRESCENCES

Whitehead himself explained it: the word 'concrescence' is derived from the common Latin verb 'concrescere', which means 'to grow together'. It also has the advantage that the adjective 'concret' is commonly used to refer to an accomplished physical reality. Thus, the word concrescence is useful to translate the

---

8   Whitehead, *Adventures of Ideas*, 280.
9   H. Bergson, *L'Evolution Créatrice*, trans. Arthur Mitchell (New York: Henry Holt & Co., 1907/1911), 7 and 15.

notion of a plurality of things acquiring a complex but completed unity. But it does not account for the underlying creative novelty. In fact, Whitehead applies this word and the concept it represents more particularly to moments of global crystallisation, those that concern the whole world: 'Concrescence is the name for the process in which the universe of many things acquires an individual unity in a determinate relegation of each item of the "many" to its subordination in the constitution of the novel "one".'[10]

As Whitehead repeated, concrescence itself does not take place in time. Yet Whitehead distinguishes several phases in the concrescence, but these are logical and not temporal. As John Cobb explains in his *Whitehead Word Book*, even if a concrescence occurs, temporally speaking, in a single block, in order to understand it one must analyse the stages or phases of its occurrence. In fact, Whitehead has devoted a large part of *Process and Reality* to a long explanation of these phases, which we call logical, since they are not temporal; Whitehead, on the other hand, prefers to call them *the genetic analyses* of the occasion. In his 1926 lecture on Time, he distinguishes three stages in each replacement of concrete reality by another: new reality replaces old reality, it is replaced by another reality, but it is also in itself a process of replacement. The concrescences are crystallisations. They happen and perish, but these happenings do not unfold time. Time comes from something else. To understand this, we must remember that time as we understand it, the common notion so called, is continuous, and is an abstraction from the *process*, the real motor of changes in the world and of evolution. Our time is a kind of envelope flying over concrescences, neglecting the accidents they form, a bit like we construct the trajectory of an electron in the photographic

---

10 Whitehead, *Process and Reality*, 211.

plates of a Wilson's chamber by interpolating a continuous line through the droplets of cloud. Or more exactly, let's say that each concrescence brings with it a brick of time, and our spirit articulates, concatenates these bricks, weaving continuous time. The time of things comes from the continuity of the immersion of past things in the present concrescences. Man's time comes from the integration of concrescences in memory, it is an abstraction, a way of surfing over the waves formed by successive concrescences.

ACTUAL OCCASIONS

Each concrescence gives rise to an *actual entity*, also called an *actual occasion*. This last name may seem very strange today. Why not call it 'thing' or 'object'? Whitehead tells us:

> The most general term, 'thing'—or, equivalently, 'entity'—means nothing else than to be one of the 'many' which find their niches in each instance of concrescence. Each instance of concrescence *is itself* the novel individual 'thing' in question. There are not, 'the concrescence' *and* 'the novel thing': when we analyze the novel thing we find nothing but the concrescence.[11]

Therefore, 'an actual occasion is nothing but the unity to be ascribed to a particular instance of concrescence. This concrescence is thus nothing else than the "real internal constitution" of the actual occasion in question'. Whitehead no doubt drew this word of *actual occasion* from the seventeenth-century authors with whom he was familiar. What is probable is that he uses it to underline the contingent character of reality. The

---

11 Ibid., 211.

word *occasion* does indeed have that tone, doesn't it? When we say that we have had the occasion to meet someone or to go to a certain place, we mean that we have seized a chance. In any case, the most important fact for Whitehead in the concrescence of an actual occasion is the creativity by which the plurality of occasions reaches concrete unity. For this, transition to unity is never fully determined. It is a bit like assembling marquetry to form a parquet floor or a Roman mosaic for the reception room of an imperial palace. All the available pieces, the partial concrescences of the past, do not necessarily contribute to the assembly; for a piece to participate, its shape or its colours must contribute to the general harmony.

> Thus we can never survey the actual world except from the standpoint of an immediate concrescence. [...] The creativity in virtue of which any relative complete actual world is, by the nature of things, the datum for a new concrescence is termed *transition*. Thus, by reason of transition, *the actual world* is always a relative term, and refers to that basis of presupposed actual occasions which is a datum for the novel concrescence.[12]

An actual entity thus has a triple character: (*a*) that which the past 'gives' to it, the way in which it integrates the past and which gives it the continuity that we observe. This will be developed in the following chapter; (*b*) the one that is aimed at in its process of concrescence, and which Whitehead qualifies as 'subjective'. It is basically about the demand for harmony, or perhaps better yet, the *thirst for harmony* that Whitehead believes he discerns in each act of concrescence and more generally in the unfolding of the world; (*c*) that of the final value associated

---

12   Ibid., 211.

with the concrescence achieved, which Whitehead relates to the 'transcendent' creativity manifested by this concrescence.

At any given moment, the world around us is thus made up of a multitude of actual entities, each one of them produced by an act of concrescence. And the world itself as it is given to us is an actual entity, the product of a global concrescence. However, Whitehead wishes to emphasise that in his opinion these levels of concrescences deserve to be distinguished: microscopic concrescences are not quite the same as macroscopic or global concrescences:

> To sum up: There are two species of process, macroscopic process, and microscopic process. The macroscopic process is the transition from attained actuality to actuality in attainment; while the microscopic process is the conversion of conditions which are merely real into determinate actuality. [...] The former process is efficient; the latter process is teleological. [...] The notion of 'organism' is combined with that of 'process' in a twofold manner. The community of actual things is an organism; but it is not a static organism. It is an incompletion *in process* of production. Thus the expansion of the universe in respect to actual things is the first meaning of 'process'; and the universe in any stage of its expansion is the first meaning of 'organism'.[13]

Whitehead insists on the fact that when we talk about an event, an actual entity in the world, we must always think of them as a *process taking place in an organism*, with its characteristic phases. This process reproduces at the level of our environment what the macrocosm of the universe is: a process

---

13 Ibid., 214.

that continues from phase to phase, each phase serving as a real support for the subsequent phase.

This continuous chain of actualisations within a global organism explains why, for Whitehead, in a sense the future enjoys an objective reality in the present, although it cannot yet be qualified as a formal actuality or as a concrete reality. 'For it is inherent in the constitution of the immediate, present actuality that a future will supersede it. Also conditions to which that future must conform, including real relationships to the present, are really objective in the immediate actuality', in as much as the constraints of causality, represented by all the past occasions with which the present occasion weaves prehensions, surround it and penetrate its constitution. Moreover, this enclosing, this interpenetration of actual occasions makes it possible to say that each of them does not really die, but experiences their objective immortality.

In another passage of his book, Whitehead comments: 'The "objectifications" of actual entities in the actual world, relative to a definite actual entity, constitute the efficient causes out of which *that* actual entities arises; the "subjective aim" at "satisfaction" constitutes the final cause, or lure, whereby there is determinate concrescence; and that attained "satisfaction" remains as an element in the content of creative purpose. There is, in this way, transcendence of the creativity; and this transcendence effects determinate objectifications for the renewal of the process in the concrescence of actualities'.[14]

SATISFACTION

The concretisation of the occasion is completed. Satisfaction marks the conclusion of the concrescence. Whitehead imagines

---

14   Ibid., 87.

that the thing created or recreated is pleased with the stage it has reached. In Genesis, when God created heaven and earth, he looks at his creation and says that it was good. Whitehead somehow secularises this feeling by bringing it to the level of the concrescence.

'The satisfaction is merely the culmination marking the evaporation of all indetermination; so that, in respect to all modes of feeling and to all entities of the universe, the satisfied actual entity embodies a determinate attitude of "yes" or "no".' Satisfaction thus consists in reaching the particular possible ideal which constitutes, in a way, its final cause, and it is measured in proportion to the ideal of harmony that it was possible to reach:

> This process of integration of feeling proceeds until the concrete unity of feeling is obtained. In this concrete unity, all indetermination as to the realization of possibilities has been eliminated. The many entities of the universe, including those originating in the concreteness itself, find their respective rôles in this final unity. This final unity is termed the "satisfaction". The 'satisfaction' is the culmination of the concrescence into a completely determinate matter of fact. In any of its antecedent stages the concrescence exhibits sheer indetermination as to the nexus [the network of entities] that unites its multiple components.[15]

Thus, in the philosophy of the organism, each actual occasion, or each event, considered in itself, is nothing more than a passage between two ideal completions, namely, the passage from its components in their diversity to these same components in their concrete being-together. Whitehead sees two possible

---

15 Ibid., 211.

doctrines concerning this passage. The first is that it is an external Creator who deliberately chooses this final whole-being from nothing. For the second, there is a metaphysical principle inherent in the nature of things such that this passage occurs essentially automatically, so that we see nothing in the Universe but concrete examples of such passages with their components. If one adopts the latter doctrine, Whitehead says, then the word 'creativity' expresses nothing more than the notion that every event is a transition producing novelty. 'If guarded in the phrases Immanent Creativity, or Self-Creativity, it avoids the implication of a transcendent Creator',[16] the author concedes. But, as will be seen in the last chapter, Whitehead will ultimately prefer to call this universal agent of creativity 'God', even though he does not consider it the personal God of the monotheistic religions but rather as an entity halfway from Spinoza's *Deus sive natura* (though not to be confused with it).

APPEARANCE AND REALITY

Whitehead was very conscious of the fact that the human brain is not designed to give a completely accurate view of reality. As is well known, Kant put a lot of emphasis on this, presenting the perceived world as a world reconstructed by thought. But Whitehead's philosophy reverses the Kantian perspective. In Kant's view, the world comes from the subject, whereas in the philosophy of the organism, the subject comes from the world. Neuroscience as it has developed today largely confirms this: the vision of the world that the brain offers us is a world interpreted, coloured by our memories and constructed to refine our subjective evaluations about our perceptions' causes

---

16 Whitehead, *Adventures of Ideas*, 236.

and prepare our actions accordingly.[17] Whitehead explains that, when we consider the process that forms the basis of an occasion of experience, the perception of persistent individualities is necessarily part of the final appearance in which the occasion finds its completion. In the first phase of the concrescence in question, the past initiates the process by virtue of the energy it brings to the occasion. This is the reality from which the new occasion arises. And it is also at this moment that the world as perceived by an observer, and which Whitehead calls its *appearance*, can arise:

> There finally emerges the Appearance, which is the transformed Reality after synthesis with the conceptual valuations. The Appearance is a simplification by a process of emphasis and combination. Thus the enduring individuals, with their wealth of emotional significance, appear in the foreground. In the background there lie a mass of undistinguished occasions providing the environment with its vague emotional tone. In a general sense, the Appearance is a work of Art, elicited from the primary Reality.[18]

CONCRETENESS AND CONTEMPORANEITY

The world of concrescences does not necessarily imply that an observer is there to capture it. Whitehead is not likely to be accused of solipsism—the theory that the outside world has no real existence, because we are only sure of our own existence—or of sensualism, in the sense that reality would not exist without consciousness to apprehend it. But when the observers are there, they are necessarily part of the concrescence, and the relations that minds weave with other concrescences,

---

17 R. Lestienne, *Le Cerveau Cognitif* (Paris: CNRS Editions, 2016).
18 Whitehead, *Adventures of Ideas*, 281.

with the objects they identify, are necessarily part of the global concrescence. From this point of view, in the philosophy of the organism, there is never a clear distinction between subject and object. The world as perceived, the Whiteheadian appearance, is a particular aspect of the world, this global organism.

This is why Whitehead gives great importance to the notion of contemporaneity. As we know, this common notion was revised by the discovery of the finitude of the speed of light, and then of causal relations, in Einstein's theory of relativity. For Whitehead, two events are contemporary when they occur in causal independence from each other. Thus, two contemporary occasions are such that neither belongs to the past cone of the other. In other words, these two occasions are not directly related to the same efficient cause.

> The vast causal independence of contemporary occasions is the preservative of the elbow-room within the Universe. It provides each actuality with a welcome environment for irresponsibility. 'Am I my brother's keeper?' expresses one of the earliest gestures of self-consciousness. Our claim for freedom is rooted in our relationship to our contemporary environment.[19]

But, on the other hand, Whitehead naturally recognises that two contemporary opportunities may have their origin in a common past, and their future may have a common aspect. 'Thus, indirectly, *via* the immanence of the past and the immanence of the future, the occasions are connected. But the immediate activity of self-creation is separate and private, so far as contemporaries are concerned'.

The physicist reader will have recognised in Whitehead's comments on contemporaneity the influence of the development of the

---

19 Ibid., 195.

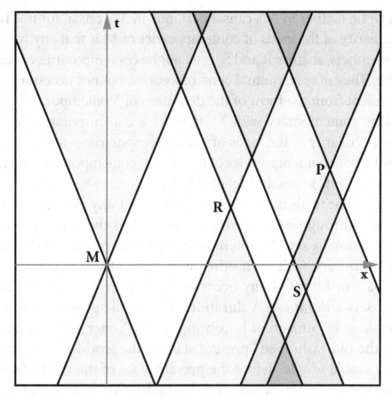

Figure 5.1. *The particle-events M, P, R, S each have their own light-cones, which determine for each their past, their future and their 'elsewhere'. These overlapping regions define the possible causal relationships between each. Note that both P and S are contemporary of M, but not mutually contemporary.*

Einsteinian theory of relativity. The latter has largely emphasised the light-cones, which separates from an original event the different regions of space-time: the past, the future, and the elsewhere. In Figure 5.1, the light-cones of the events or actual occasions M, R, S, and P are shaded. The slope of the light-cones is fixed by the speed of light, which is, as we know, the limiting speed of propagation of causal actions. This figure is inspired by Whitehead's own comments on the notion of contemporaneity: 'Actual occasions R and S, which are *contemporary* with M, are those actual occasions which lie neither in M's causal past, nor in M's causal future. The peculiarity of the locus of contemporaries of M is that any two of its members, such as R and S, *need not* be contemporaries of each other. They *may* be mutual contemporaries, but not necessarily. It is evident from the form of the definition of "contemporary", that if R be contemporary with M, then M is contemporary with R. This peculiarity of the locus of M's contemporaries—that R and S may be both contemporaries of M, but not contemporaries of each other—points to another set of loci'.[20]

Whitehead thus invites us to note that any inclined plane passing through M and whose slope is less than the speed of light defines a set of contemporary present occasions that are contemporary with each other, and none of which belongs to the past or future of any occasion in that set. Whitehead calls such sets a *duration*. 'A duration is a complete locus of actual occasions in "unison of becoming", or in "concrescent unison". It is the old-fashioned "present state of the world"'. So there is not a single way to define the present state of the world for an observer or an actual occasion located in M, there is not only one 'duration' associated with this point, but an infinity of them.

In the philosophy of the organism, the present state of the world is a multitude of concrescences that are independent of each other, and therefore in each of which there is a place for the advent of a dose of novelty. But it is also a whole that depends solidly on its past, on this infinitely dense fabric of previous concrescences. These anterior concrescences give to certain evolutionary lines between concrescences qualities of permanence, which an observer will naturally designate as 'perennial things' or 'permanent objects'. What then is the status of causality in the philosophy of the organism? This is what we must now examine.

---

20 Whitehead, *Process and Reality*, 320.

CHAPTER 6

Permanence of Objects, Causality, and Bifurcations of the Becoming

> Causation is nothing else than one outcome of the principle that every actual entity has to house *its* actual world.
>
> A. N. Whitehead, *Process and Reality*, 1929, 80

Alfred Whitehead depicts the present that is offered to us as a collection of actual entities, and as a global actual entity, each of them being the result of a concrescence, of the sudden crystallisation of a potential that was there but awaiting its actualisation. As Alex Parmentier writes, 'The actual entity is formally a "pulse of actuality", a "drop of experience", indivisibly one and having a unique individual character'.[1]

The content of these concrescences, of course, is not the result of chance: it depends essentially on past actualisations,

---

1   A. Parmentier, *La philosophie de Whitehead et le Problème de Dieu* (Paris: Beauchesnes, 1968), 222.

as the whole development of classical science teaches us, which the latter formalises through the determinism of the equations of evolution. In Whitehead's system, it depends on the 'prehensions' that the actual occasion in the process of concretisation projects towards other actual occasions, especially those of the immediate past. What are these 'prehensions'? What are the mechanisms by which the links between the various concrescences are woven? Whitehead is particularly inspired here by Leibniz and his theory of 'monads'. Let us recall that Leibniz calls thus the constitutive elements of the world, which are for him these kinds of bubbles or metaphysical atoms, because they are without extent (one can think of the elements of a conceptual space, such as the wave functions of wave mechanics). But like soap bubbles that reflect the outside world on their surface, monads have the property of 'feeling' the other monads that populate the world, and these 'feelings' indeed weave the fabric of the world. Because of these 'feelings', 'perceptions', and 'apperceptions', each Leibnizian monad not only represents the state of an element of the world but is itself a kind of condensed world. Whitehead concludes this tribute to Leibniz by stating that, if his theory is indeed a theory of monads in the line of this author, it differs from it. For Leibniz, writes Whitehead, 'monads change. In the organic theory, they merely *become*. Each monadic creature is a mode of the process of "feeling" the world, of housing the world in one unit of complex feeling, in every way determinate. Such a unit is an actual occasion; it is the ultimate creature derivative from the creative process'.[2] Whitehead, concerned with the terminology, criticises the Leibnizian lexicon, mixed with notions of consciousness and representation. However, following the model of the Leibnizian

---

2   A. N. Whitehead, *Process and Reality* (New York: Macmillan, 1929), 80.

term 'apprehension' meaning complete understanding, he 'uses the term *prehension* for the general way in which the occasion of experience can include, as part of its own essence, any other entity, whether another occasion of experience or an entity of another type. This term is devoid of suggestion either of consciousness or of representative perception'.[3]

THE PREHENSIONS

The word 'prehension' first appears in Whitehead's *Science and the Modern World* (1925) where it was introduced during a long discussion of human perception as the mechanism by which we appropriate through sight objects that we distinguish and recognise and which are therefore 'grasped' in consciousness. But Whitehead, in his philosophy of the organism, wants to extend this capacity to any actual entity, conscious or not, and therefore adopts the word prehension as a generalisation of the idea of apprehension, concerning thinking or non-thinking entities, and therefore as a synonym of 'uncognitive apprehension'.[4] Let's take a hydrogen atom; we know that it consists of two particles: a proton, relatively heavy and positively charged, and a much lighter electron, negatively charged. The proton holds the electron around it by the electromagnetic force that is exerted between opposite electric charges. Let us distinguish in it these two constituent particles: the proton and the peripheral

---

3   A. N. Whitehead, *Adventures of Ideas* (New York: Free Press, 1933), 234. Whitehead is wary of the terms 'consciousness' and 'perceptual representation' because for the philosophy of the organism any actual occasion, whether a living being or not, has 'feelings' about other actual occasions that are not necessarily conscious. For him, it is not consciousness, but experience that is primary.

4   A. N. Whitehead, *Science and the Modern World* (New York: Macmillan, 1925), 70.

electron, which, like Whitehead, we will consider as two constantly updated actual occasions. At each concrescence of this system, proton and electron prehend each other; this is the Whiteheadian vision of the electromagnetic force.

The theory of prehensions occupies a large part of Whitehead's master work, *Process and Reality*. Victor Lowe, Whitehead's former student and biographer, explains that this is where Whitehead overtakes Bergson in their respective analyses of the notion of time. While Whitehead develops the idea of process and his elaborate theory of prehensions, 'Bergson, believing theory of no avail, uses poetic imagery to supplement his references to "melting" and "interpenetration"'. Both men hold that the true process [the concrescence] is indivisible; but for Whitehead, it always has the shape of an analysable concrescence, whereas the issue of Bergson's meditation was the intuition of 'pure, unadulterated inner continuity (duration), continuity which was neither unity nor multiplicity'.[5] For Whitehead, the continuity of the flow of experience is a superficial property that is only made evident through the integrative properties of consciousness, whereas the underlying reality is a succession of 'drops of experience'. For Bergson, on the contrary, 'continuity is the fundamental fact, and there are no drops in Whitehead's sense, but only static states artificially abstracted by our acts of attention or by psychological analysis'.[6] But, in fact, Whitehead goes further; he develops the idea of prehensions as a mechanism by which successive concretenesses both incorporate the past and account for causality, but also leave room for choices, for potentialities that can offer alternative paths to the future of the world.

---

5   V. Lowe, *Understanding Whitehead* (Baltimore: The John Hopkins Press, 1962), 260.
6   Ibid.

For the content of the concrescences, while largely determined by past concrescences, is not entirely determined by them. This is, without a doubt, a great originality of the philosophy of the organism proposed by our author. There is in this global organism, that is, in the world, a principle of harmony, an emancipating principle, which allows the appearance of novelty and gives the actual occasion a 'plus' with respect to previous actual occasions. 'There are two principles inherent in the very nature of things, recurring in some particular embodiments whatever field we explore—the spirit of change, and the spirit of conservation. There can be nothing real without both. Mere change without conservation is a passage from nothing to nothing. [...] Mere conservation without change cannot conserve. For after all, there is a flux of circumstance, and the freshness of being evaporates under mere repetition.'[7]

Amongst the prehensions, there are therefore physical forces, but we will see later that in Whitehead the notion of prehension is broader than that. Alex Parmentier writes that 'the analysis of a prehension reveals three factors, which clearly demonstrate its relational nature: The subject who prehends, the datum which is prehended, and the subjective form: *how* the subject prehends this datum'.[8]

The prehending subject is the actual occasion in the process of concrescence; it selectively prehends the actual occasions in its environment. To the extent that the prehensions represent causal links, these prehensions cannot be instantaneous, as we have already discussed in the previous chapter. This does not pose a difficulty, provided that the past concrescences to be prehended are separated from the concrescence in the process of becoming in question by a sufficient amount of accumulated

---

7   Whitehead, *Science and the Modern World*, 201.
8   Parmentier, *La philosophie de Whitehead et le Problème de Dieu*, 219.

time. The datum that is prehended is the actual occasions with which the subject exchanges prehensions, which are not necessarily limited to the immediate vicinity of the subject: the universal attraction of occasions endowed with mass testifies to prehensions that know no limits. Thus, following the example of the Leibnizian monads, each current entity weaves definite links with all those who have woven the history of the universe.[9]

The subjective form of the prehension is undoubtedly a specificity of Whitehead, linked to his idea of the 'mental pole' of each actual occasion. In order to best conform to the creative principle of harmony and thereby achieve optimal 'satisfaction', Whitehead teaches that the actual occasion in the process of concrescence can select the actual occasions with which it is going to link itself by prehensions, 'accepting' some and 'rejecting' others, and even to draw from a virtual reserve of which we will speak later, the 'eternal entities', sometimes also called 'conceptual entities', of visibly Platonic inspiration, but whose existence proves to be essential to Whitehead's deist philosophy, as we will discuss later.

For the moment, let us retain thus that the prehensions make it possible to account at once for the causality, the permanence of the objects, and the creativity in nature. Causality, because the actual occasion in the process of concrescence 'prehends' a variable number of previous concrescences; permanence of objects, because of the similarity between successive concrescences that define a natural lineage of concrescences, and thus make it possible to speak of a permanent object; and finally creativity of the *process*, because of the principle of harmony mentioned above, which gives the concrescence the possibility of selecting the actual occasions with which it will exchange prehensions. As John Cobb points out:

---

9   Ibid., 221.

Nevertheless, no occasion is simply the reenactment of the past. One reason is that even if an occasion in an enduring object [...] derives much of its content from its immediate predecessor, it is also prehending other occasions. It takes account of its entire actual world, and that cannot be exactly the same as that of its predecessor. The new occasion must integrate what it receives from many sources, but it can do that only if it appropriates those sources selectively. This applies even to its predecessor in the enduring object. Thus even conformation to the past introduces a measure of novelty.[10]

The end result of such a process is the objectification of today's world. 'Objectification is an operation of mutually adjusted abstraction, or elimination, whereby the many occasions of the actual world become one complex datum'.[11]

CLASSICAL CAUSALITY AND TEMPORAL IRREVERSIBILITY

I have mentioned that for Whitehead, in the act of concrescence of an actual occasion, many previous actual occasions are 'prehended', that is, assimilated, incorporated, memorised, immersed in the actual occasion. This is, in Whitehead, the mechanism equivalent to the cumulative character of time in Bergson. Whitehead describes it as 'physical memory' and it is for him the means to account for physical causality. In his 1926 article "Time", our author imagines two immediately subsequent occasions A and B, that is, occasion B occurs after occasion A to which it is closely linked by a strong prehension.

---

10 J. Cobb, *Whitehead Word Book, a Glossary with Alphabetical Index to Technical Terms in Process and Reality* (Claremont: P&F, 2008), 61.
11 Whitehead, *Process and Reality*, 210.

The full transaction between A and B, consisting of the pair of objectivations, constitutes A and B as poles in a linkage. A, in its function of a constituent member of this linkage, A and B, is more complete than A in abstraction from the linkage. For the indetermination of B in A, which clings to A in abstraction, is removed by the actual concretion of B in the full linkage. Thus, in the community of A and B, the incompleteness of A by reason of B is rectified by the completion B so far as its transaction with A is concerned: A has thereby an added meaning. Hence, each occasion A is immortal throughout its future. For B enshrines the memory of A in its own concretion, and its essence has to conform to its memories. Thus physical memory *is* causation, and causation *is* objective immortality.[12]

But it is not only the passage from event A to event B that constitutes causality. It is the whole set of past occasions that is taken into account in the concrescence of the actual entity B. 'No event can be wholly and solely the cause of another event', Whitehead says in *Modes of Thought*. 'The whole antecedent world conspires to produce a new occasion. The only intelligible doctrine of causation is founded on the doctrine of immanence. Each occasion presupposes the antecedent world as active in its own nature'.[13]

This way of seeing the present, not as the advent of a brand-new reality, but as the nesting, or better perhaps the 'telescoping' of the past into the present, making the present ever richer and the past immortal, was perhaps inspired not only by Bergson.

12 A. N. Whitehead, "Time." (1926). Lecture reproduced in: *The Interpretation of Science, Selected Essays* (Indianapolis: Bobbs-Merrill Co., 1961), 243.

13 A. N. Whitehead, *Modes of Thought* (New York: Macmillan, 1938), 225–26.

Should we see in this search for a kind of immortality of things and disappeared beings a reconciliation with the world, initiated by the disappearance of his son Eric? Things are undoubtedly more complex...

Whitehead will develop his presentation of causality as efficient cause in *Process and Reality*. There, he explains that causality is nothing more than the expression of the transition from one actual entity to another. The reason why the cause is objectively present in the effect is that the prehension between cause and effect is inseparable from the past occasion that participates in the concretisation of the actual occasion. 'This passage of the cause into the effect is the cumulative character of time. The irreversibility of time depends on this character.'[14] Thus, in causality as understood in the philosophy of the organism, operations are directed from antecedent organisms to the target organism. This is efficient causality as brought to light by classical mechanics, from Descartes to Newton, and specified by the deterministic equations of motion.

THE PERMANENCE OF OBJECTS

'Our lives are dominated by enduring things', writes Whitehead, 'each experienced as a unity of many occasions bound together by the power of inheritance'. This is such a banality that we pay no attention to it. But in what anguish would we live if in an ever-changing world we could not distinguish those permanencies that allow us to live by appropriating them, by acting on them? Whitehead sees them as the first conquest of the natural evolution of the Universe, starting from the original chaos:

---

14 Whitehead, *Process and Reality*, 237.

Each such individual endurance collects into its unity the shifting qualities of its many occasions. Perhaps it is the thing we love, or perhaps it is the thing we hate. There is a bare 'It'—a real fact of the past, stretching into the present, which concentrates upon itself the wealth of emotion derived from its many occasions. Such enduring individualities, as factors in experience, control a wealth of feeling, an amplitude of purpose, and a regulative power to subdue into the background the residue of things belonging to the immensity of the past. Surely, this is what Descartes must have meant by the *realitas objectiva* which, according to his doctrine, clings more or less to our perceptions.[15]

The persistence of the objects is thus explained by the reiteration of certain characters, in the course of a chain of successive concrescences of actual entities which prehend each other; these common characters make manifest a structure which reproduces itself exactly (or which slowly modifies itself) at each concrescence; it is that which gives rise to the notion of permanent object. Whitehead remarks that, when we take matter as a foundation, the property of permanence is necessarily regarded as an arbitrary property at the basis of the order of nature; but when we take the organism as a foundation, we can conceive this property as the result of evolution.[16] As the concrescences succeed one another, structures endowed with this property of stability have appeared from the original chaos, more and more varied according to the scale of their size and complexity.

Finally, it should be noted that when analysing the property of persistence within the framework of the philosophy of the organism—in which successive concrescences not only bind

---

15 Whitehead, *Adventures of Ideas*, 280.
16 Whitehead, *Science and the Modern World*, 111.

together the successive concrescence as in a pearl necklace but absorb or embed them into each other—the permanence of objects is not a persistence beyond oneself, but truly 'persistence within oneself', in Whitehead's own words. The permanence of objects thus takes on a particular depth here, which may explain the attraction of other philosophers and scientists to the notion of substance, which Whitehead, however, rejects.

'FEELINGS' AND PHYSICAL LAWS

An actual occasion in the process of concrescence prehends a wide variety of actual occasions in its environment, but not indistinctly all of them: some connections are excluded. Whitehead generally uses the word 'feeling' to refer to positive prehensions, as opposed to those rejected. Feelings are thus the prehensions between actual entities that are really taken into account in the concrescence of the actual occasion being realised. In this way, 'in positive prehensions the given "datum" is preserved as part of the final complex object which "satisfies" the process of self-formation and thereby completes the occasion'.[17] Moreover, 'simple feelings' are those that link the occasion in the process of concrescence to physical occasions, as opposed to the Platonic 'eternal entities' that we have not yet introduced. Basically, then, 'simple feelings' are the forces at play in nature: electromagnetic force between charged particles, gravitational force, and nuclear forces:

> A simple physical feeling has the dual character of being the cause's feeling re-enacted for the effect as subject. [...] By reason of this duplicity in a simple feeling there is a vector character which transfers the cause into the effect. It is

17 Whitehead, *Adventures of Ideas*, 234.

a feeling *from* the cause which acquires the subjectivity of the new effect without loss of its original subjectivity in the cause. Simple physical feelings embody the reproductive character of nature, and also the objective immortality of the past. In virtue of these feelings time is the conformation of the immediate present to the past. [...] Physical science is the science investigating spatio-temporal and quantitative characteristics of simple physical feelings.[18]

Naturally, the word 'feeling', even more than the word 'prehension', evokes a subjectivity of the actual occasion. This is indeed what Whitehead wishes to make us understand. Even at the level of inert matter, even at the level of elementary particles, there is a subjectivity of actual occasions. Certainly, the forces of nature are universal, but a neutron, which carries no electric charge, does not prehend an electron from the perspective of the electric field. More generally, the composition of the palette of feelings depends on the end pursued by the actual occasion in concrescence, which Whitehead describes as 'satisfaction', that is, the highest possible degree of harmony, of functionality of the organism in which the occasion is immersed—first the local environment, and, gradually, nature, in general.

What is the source of the idea of this subjectivity adopted by Whitehead? On this subject, Victor Lowe tells an interesting anecdote in *Understanding Whitehead*. During a conversation held in August 1942, Whitehead confided to him that the contemporary philosopher who had influenced him most was Samuel Alexander, because he and Alexander 'conceived the problem of metaphysics in the same way'. In particular, both thought that the idea of the unity of the universe (Spinoza) and the multiple individuals (Leibniz) had to be reconciled.

---

18 Whitehead, *Process and Reality*, 237–38.

Moreover, Alexander, almost alone amongst Whitehead's contemporaries, did not seem to assume, no more than Whitehead himself, that experience be fundamentally an experience through sensory data, but rather something akin to what Bergson calls 'intuition'.[19] Indeed, these reflections are essentially confirmed by Whitehead himself in *Process and Reality*.[20]

Physicists may not feel comfortable with this anthropomorphic language, although Whitehead asks that he be understood and denies to be a panpsychist; indeed, he repeatedly asserts that what he calls 'the mental pole' of actual occasions has nothing to do with consciousness, because 'experience' is the condition for the occurrence of 'consciousness', not the other way around. This means that inanimate actual occasions can have the experience of the external world, even 'feel' it, with the emotional connotation that this word implies, without being endowed with consciousness. 'We must admit that it is difficult to conceive of a "feeling" that is not a "conscious feeling,"' acknowledges the physicist Henry Stapp, 'for the latter is the only kind of feeling that we actually know, or know of. But if we accept that our conscious feelings are complex versions of simpler elements that can act dynamically upon other like elements and merge with them to form more complex elements of the same kind, then we have, I think, gained an important insight into what Whitehead was driving at with his choice of word. And we will have established a basis for understanding how consciousness can emerge from realities that are not conscious'.[21] For example, for him, in the human body, one cell

---

19 Lowe, *Understanding Whiteheads*, 264. We have already met in Chapter 3 the English philosopher Samuel Alexander, the author in particular of *Space, Time and Deity*.
20 Whitehead, *Process and Reality*, 41.
21 H. Stapp, "Whitehead, James, and the Ontology of Quantum Theory," *Mind and Matter* 5 (2007): 83–109, 96.

'feels' the other cells, not only of its immediate surroundings but of the whole society that forms the total organism. In the Whitehead's language:

> The feelings are inseparable from the end at which they aim; and this end is the feeler. The feelings aim at the feeler, as their final cause. The feelings are what they are in order that their subject may be what it is. [...] An actual entity feels as it does feel in order to be the actual entity which it is. In this way an actual entity satisfies Spinoza's notion of substance; it is *causa sui*. The creativity is not an external agency with its own ulterior purposes. All actual entities share with God this characteristic of self-causation.[22]

CONCEPTUAL PREHENSIONS: THE UNCERTAIN FUTURE

Again, however, the concretisation of the actual organism does not obey solely to efficient causality. There is a part of concrescence that does not depend on efficient causes but on a true final cause. What is it? For Whitehead, it is this principle of harmony that we have been talking about and towards which all actual occasions tend. Accordingly, each actual occasion, each concrescence has a subjective and creative character. 'There is the becoming of the datum, which is to be found in the past of the world; and there is the becoming of the immediate self from the datum. This latter becoming is the immediate actual process. An actual entity is at once the product of the efficient

---

22 Whitehead, *Process and Reality*, 222.

past, and is also, in Spinoza's phrase, *causa sui*,[23] for it builds itself by adding to the given what will enrich the world.

Moreover, as we have seen, in the philosophy of the organism, a dose of subjectivity is granted to any actual entity, and not only to thinking beings. Whitehead notes that, while Descartes posits the thinker as creating thought, the philosophy of the organism reverses this order and conceives of thought as a constituent operation in the creation of the occasional thinker. There is a 'mental pole' in everything. This is not, however, pure panpsychism, inasmuch as this mental pole, at least in inanimate nature, is obviously not endowed with a consciousness or even with a self-reflective capacity, but with a simple project, that of realising itself as harmoniously as possible in the concert of actual occasions that form the world, and thus achieving what Whitehead calls 'full satisfaction'.

Let us sum up, at the risk of repeating ourselves. Simple feelings of causality form the foundation of physical causality. The primary activity of each occasion is to absorb the actual occasions of the past, especially those of its environment, with which it is directly related. But this aspect is far from totally constituting the actual entity in its act of concrescence. For, thanks to its mental pole, the actual occasion is capable of bringing its own stone to its edification, to its concretisation, to its concrescence. 'Its mental side is its own creativeness, its desire for and realisations of ideal forms (including its own terminal pattern)' allows it to form a new unified reaction to its antecedents. For this purpose, the occasion can not only select the present occasions with which it weaves prehensions but also select eternal entities, prehend them, and draw from them the source of their

---

23 Ibid., 150.

inventiveness. 'Each occasion is a fusion of the already actual and the ideal'.[24]

What are these eternal entities? For Whitehead, they exist and are always available for weaving prehensions with actual occasions, although they always remain in the realm of potentiality. They form the infinite nuances of the possible. Whitehead explains that any eternal object can be positively integrated into any given concrescence, if the latter selects it by a feeling, just as it can be excluded by a negative prehension. For the actual entity is a stubborn fact, but an eternal object never loses its 'accent' of potentiality.[25] There is, as already said, in the idea of eternal entities a reminiscence of Plato's Eternal Ideas, for example, the ideals of Beauty or Good. But in Whitehead it goes further. In a sense, eternal (or conceptual) entities together constitute, in part, what Whitehead calls 'the primordial nature of God'.

It is here that one encounters, for the first time, this transcendent, but for him impersonal principle, which he calls God. It will be found again and again, and will be discussed in the final chapter of this book. Naturally, the God of Whitehead is closely associated with the principle of creative harmony, of creativity, which he sees at work in nature, at each concrescence, with its two poles of causality and creativity. 'God is the principle of concretion; namely, he is that actual entity from which its self-causation starts. [...] If we prefer the phraseology, we can say that God and the actual world jointly constitute the character of the creativity for the initial phase of the novel concrescence'.[26]

Through the mix of causality, creativity, and finality that Whitehead introduces into the *process*, the Whitehead's concept

---

24 Lowe, *Understanding Whitehead*, 43.
25 Whitehead, *Process and Reality*, 239.
26 Ibid., 244–45.

has obviously broken with the contemporary science that it was fed with. But there is another reason that pushed him to take this step, drawn from the very development of science in his time. Classical science saw change as the progressive modification of the characters of a material substance, the position of a particle in three-dimensional space for example—or more precisely, for established science, the change of a parameter of position or velocity in the six-dimensional 'configuration' space—characters that vary continuously over time. But, notes Whitehead, in 1929, this vision of the world has already changed; science and philosophy must adapt: don't modern physicists conceive of energy as transmitted, not continuously, but in packets or defined 'quanta'? From now on, Whitehead argues, science, philosophy, and cosmology must do justice to atomism, as well as to continuity; to causality but also to the specific properties of organisms, such as memory and perception. 'But so far there has been no reference to the ultimate vibratory characters of organisms and the "potential" element in nature.'[27]

---

27 Ibid., 239.

CHAPTER 7

World Solidarity and Quantum Entanglement

Let us now turn to the possible relations between metaphysics and some important developments in contemporary physics, and first of all in quantum physics. Although quantum mechanics developed in the very years when Whitehead was maturing his philosophy of the organism, we do not know of any direct contact between the founders of this theory and the Harvard philosopher. However, we have already seen that Whitehead was attentive to the developments of wave mechanics, when in 1925 he criticised the notion of trajectory and introduced his notion of time devoid of instants. However, at that time physicists had not yet realised that wave mechanics was causing even more severe upheavals in our notion of reality: it calls into question the classification we spontaneously make of reality into distinct objects, separated in space, each with its own properties. In short, the development of quantum mechanics taught that the notion of locality—dear to Einstein who had built his theories of relativity on it—had to be completely revised.

QUANTUM MECHANICS DENIES THE IDEA OF SPATIAL LOCALISATION

In 1927, Max Born proposed the probabilistic interpretation of wave mechanics to the Solvay Congress in Como: the fundamental equation of quantum mechanics (Schrödinger's equation) describes the behaviour of the wave associated in this theory with a particle, and not the behaviour of the particle itself. The wave is not material: it only indicates the probability that the studied particle will appear at a given location, if one tries to locate it at a given instant.

This proposition immediately aroused Einstein's mistrust. As I have already recounted in *The Children of Time*, Einstein responded to Born by proposing the following thought experiment. Let us imagine the movement of a particle moving from left to right, which is subject to passing through a small diaphragm placed perpendicularly on its path. Let us further imagine that we have, at some distance to the right of the diaphragm, a scintillation screen or a photographic plate on which the particle, after passing through the diaphragm, will come to mark its imprint. The calculation indicates that the probability wave representing the particle at the exit of the diaphragm is a divergent spherical wave propagating in all the space to the right of the diaphragm (phenomenon of diffraction). According to Born, the intensity of this wave at any point of the materialisation screen represents the probability of observing the particle at this point. Let us choose to give the detector screen a hemispherical shape, centred on the opening of the diaphragm, like a bowl in charge of collecting the diffracted particle. In this case, the probability wave will thus reach all the points of the screen simultaneously. How to understand then that the particle always materialises in only a single point, and not in two, ten, or twenty points?

Shouldn't we consider an instantaneous influence, and thus faster than the speed of light that 'warns' the other possible points of impact that the particle materialised in such a place and should therefore not appear elsewhere?

In his argumentation, Einstein actually admitted that the particle is necessarily somewhere, all along its path between the diaphragm and the screen, although quantum mechanics and its 'probability wave' are unable to account for this. Quantum physicists only accept to talk about the 'position' of the particle at the points where it is not only characterised (e.g., at the passage of the diaphragm) but actually 'measured', that is, observed concretely. Against them, Einstein claimed the right to speak of the property 'position of the particle' at any moment of its existence, even if the experimental arrangement did not allow to know it effectively.

However, Einstein's thought experiment is complicated by the inability to describe in detail the mechanism through which the particle materialises at its final position. Its impact point is the result of a complex photochemical process, so that it can be rigorously contested that the appearance of a light point on the scintillator or a silver speck on the photographic plate really reflects the property of 'position' that one would be obliged to attribute to the particle, even if the detector was not placed in its path. It is well known that a measurement process can disturb the object of observation. One cannot absolutely state that the screen only plays a passive role and that the detected position would be the same in the absence of a screen. This is why the resolution of the debate had to wait the development of more sophisticated experiments.

The development of electronic techniques in the twentieth century made it possible, as early as the 1970s, to realise an improved version of Einstein's 1927 thought experiment, contrasting the description of a photon as a wave and as a particle. Based on an idea by John Archibald Wheeler, several groups of experimenters realised a set-up in which the decision to measure either the position of a photon (pinpoint property), or on the contrary a wave property (observations of interferences between two branches of an optical bench), is taken after the photon has been introduced into the experimental set-up.[1] Let us take again here the description given in *The Children of Time* of such a barely idealised experiment, due to Abner Shimony[2] and his discussion. The photon first encounters a semi-transparent mirror, whose function is to divide the 'beam' into two parts of equal intensity. The two beams follow different paths, according to the branches A or B, but they are finally superimposed on the observation screen at the exit of the apparatus. The intensity observed on the 'O' screen is a function of the difference in path lengths between the two branches of the apparatus, a classic interference effect (Figure 7.1).

Let us first note that these interferences are well observed, even in the case where the intensity of the beams is decreased to such an extent that the photons only pass through the apparatus one by one: in this case, of course, it is by doing a statistic of the impacts on the observation screen **O** after a sufficient number of photons have passed through the apparatus that one can conclude to the existence of interferences. Each photon thus splits, in a way, into two waves propagating in the two separate branches of the interferometer before recombining on the screen. Stranger still is the result of the second part of this experiment. A fast switch (Pockels cell) is placed on the path of one of the two branches, capable of deflecting the photon, if it is present in this branch, to send it to a **P** counter that will attest to its presence on this path. This counter thus acts exactly like

---

1   One of these experiments was carried out at the Ecole Normale Supérieure de Cachan, V. Jacques et al. "Experimental Realization of Wheeler's Delayed-Choice Gedanken Experiment," *Science* 315 (2007), 966.
2   A. Shimony, "The Reality of the Quantum World," *Scientific American* 258 (1988), 46–53.

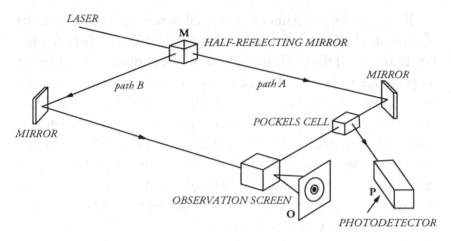

Figure 7.1. *Sketch of the experimental set-up corresponding to the single-photon beam splitting experiments, inspired by an idea by J.A. Wheeler. The quantum wave corresponding to the photon is separated into two branches by a semi-silvered mirror M. The Pockels cell placed on path A can also, in a very short time, divert the corresponding half-beam to a photodetector, where the photon is eventually absorbed and detected. When the cell is not in action, the two half-beams are recomposed into O, where the classical interference effects are observed, depending on the difference in the path between the two half-beams. When the Pockels cell is in action, the incident photon behaves as a single particle, and is detected every other time on average on the A path, by the photodetector P. The paradoxical aspect of the experiment is that these contradictory behaviours are not changed, even when the Pockels cell is activated after the incident photon has interacted with the mirror M (and before the photon has had time to propagate to the Pockels cell). Used by permission of A. Shimony, in: Lestienne R., the Children of Time (Urbana: University of Illinois Press, 1995), 93.*

a counter that would have been placed in Einstein's experiment at a precise point, somewhere between the diaphragm and the screen. The refinement of the experiment consists in the fact that the Pockels cell can be put into action after the laser beam has interacted with the semi-reflecting plate, thus after the photon has been 'split' into two beams.

Result of the experiment: if the cell is not put into action, the photon still behaves like a wave; there is interference between the two branches. If the cell is turned on, on the contrary, one observes its presence in branch **A** once every two times. But in this case, there is never a photon in branch B; the **O** detector remains mute. While there is no possible causal interaction between the materialisation of the photon in the **P** counter and its materialisation in the **O** counter, the 'decision' of the path **A** beam to behave like a particle is instantaneously understood by the path **B** beam as an order to 'vanish'. The photon thus has non local properties that concern both the branch **A** and the branch **B**.[3]

WHAT HAPPENS WHEN TWO INTERACTING QUANTUM SYSTEMS SEPARATE?

As early as 1927, Einstein thought that quantum mechanics as it had been developed, in particular by Niels Bohr and the Copenhagen School, was necessarily an incomplete theory. But

---

3   The interpretation of Wheeler's experiment according to the criterion of reality dear to Einstein (according to which any object, whose result of measurement of one or more of its properties can be predicted with certainty without interfering in any way with it, must have that property or properties of its own), can be described as follows. If in each branch of the interferometer there is a separate quantum system, and if by acting on the first system (by connecting or disconnecting the switch constituted by the Pockels cell) without acting in any way on the second one, the latter is made to manifest itself, either as a wave or as a particle, one must logically conclude that these two properties at the same time belong to this system. However, quantum mechanics denies this possibility, and asserts on the contrary that the properties of being a wave or a corpuscle are incompatible. Einstein would therefore conclude that quantum mechanics is incomplete. The resolution of the paradox requires the assertion that there is only one system in the interferometer, an inseparable photon, even if it occupies both branches at the same time.

it was only a few years later, in 1935, following extensive reflection with two of his students, Boris Podolsky and Nathan Rosen, that Einstein thought he had irrefutable proof that quantum mechanics did not provide a complete description of reality. On 4 May 1935, the *New York Times* headlined in a column on page 11 "Einstein Attacks Quantum Theory", with the subtitle "Scientist and Two Colleagues Find It Is Not 'Complete', Even Though 'Correct'". Einstein was furious. Unbeknownst to him, his young collaborator Boris Podolsky had given the newspaper an interview, revealing the subject of an article that would appear thirteen days later in one of the most serious scientific journals, *The Physical Review*. He immediately had a kind of denial published in the journal: 'Any information upon which the article "Einstein Attacks Quantum Theory" in your issue of May 4 is based was given to you without my authority. It is my invariable practice to discuss scientific matters only in the appropriate forum and I deprecate advance publication of any announcement in regard to such matters in the secular press'.

The paper published by the three researchers[4] considers two particles in a single-dimensional space that interact so that their wave function is entangled[5] and then separate. We can measure the position of the first particle, and then use the wave function of the system (which has been reduced by the measurement to a single term) to know the position of the second particle. But one could measure instead the velocity of the first particle, then use the reduced wave function (again reduced to a single term)

---

4   A. Einstein, B. Podolsky, and N. Rosen, "Can Quantum-Mechanical Description of Physical Reality Be Considered Complete?" *Physical Review* 47 (1935), 777–80.

5   Recall that the wave function of the set of two 'entangled' particles a and b is of the form $\Sigma_k c_k \psi_k \varphi_k$, where the ψ are wave functions describing the state of particle a and the φ describe the state of particle b.

of the system to know the velocity of the second particle. Since one can know the position or the velocity of this second particle without performing any measurement operation on it (which, according to the Copenhagen School, would be likely to change its state), this particle must have both a defined position and a defined velocity, a situation that wave mechanics prohibits. Thus, quantum mechanics, although it correctly predicts the measurement results, does not fully describe the properties of particles, it is incomplete. The authors suggest that the two particles have at all times in fact each a defined position and velocity, given by 'hidden variables' beyond the reach of quantum mechanics.

However, as early as 1952, a young British physicist, John Stuart Bell, began to reflect on the arguments put forward, on the one hand, by the proponents of the 'hidden variables' that one might seek to add to quantum theory in order to re-establish determinism, and, on the other hand, by the followers of the 'orthodox' interpretation of quantum mechanics, with its indeterminate or mutually incompatible magnitudes. Little by little, he understood that hidden variables attached locally to objects cannot lead in all circumstances to the same predictions as the orthodox quantum mechanics.

In 1964, he gave the first simple formulation of an argument showing that the experiment could, at least in principle, settle the hitherto philosophical debate between the proponents of separability and their opponents, on the particular example of the disintegration of a particle or the de-excitation of an atom into two particles, when the correlation between the spins[6] of these two particles is measured. Measurements made with different pairs of analyser orientations must obey, if Einstein's

---

6   The spin is the angular momentum of a particle. In the case of photons, the spin is manifested by the polarisation state of the photon, whose measurement can take two opposite values.

criterion of reality is fulfilled—if reality is always separable into distinct objects possessing properties that concern only themselves—to a mathematical inequality, which is not necessarily verified otherwise.

EXPERIMENTAL EVIDENCE OF INSEPARABILITY

At the *Institut d'Optique d'Orsay*, in the early 1980s, Alain Aspect conducted a series of experiments designed to decide between the theory of hidden variables and quantum inseparability, in the case of the production of a pair of entangled photons.[7] An analyser (polarised filter) is placed on the path of each photon and inseparability is manifested when these filters are given different orientations; for example, the polarisation axis of the first is vertical and that of the second is tilted by about 20°. The inseparability remains verified even though the orientation of the analysers is chosen after the emission of the two photons, and the distance between the two analysers is such that no signal could flow from one to the other after the passage of the first photon and before that of the second.

The conclusion of these experiments and of many similar experiments carried out since is that there is indeed in nature a fundamental inseparability, such that two systems that have interacted in the past (the two photons emitted by the same atom in the Aspect experiment) are always united, whatever their current distance. No complete description of each system is possible without reference to the other.

On the other hand, on the theoretical level, Bernard d'Espagnat had shown as early as 1975 that results contrary to the hidden variables hypothesis would go beyond the narrow framework of confirming or denying the fundamental

---

7   A. Aspect et al., *Physical Review Letters* 47 (1981), 460; 49 (1981), 91. See also *La Recherche*, 17 (1986), 1358.

principles of quantum mechanics. They would show the fundamental inseparability of nature, independently of any particular
theory[8]: in this case, *physical reality does not have the property of
locality*. D'Espagnat expressed in this way his conviction on the
necessity of abandoning the principle of separability:

> Within the framework of a realistic conception, I for my part
> see no other solution than the abandonment of the princi-
> ple of separability. This means, schematically, either that one
> must consider certain systems that are currently distant from
> one another as constituting a single system, or that between
> distant systems there are influences that are faster than light.[9]

However, this formulation raises a question. The formulation of the alternatives —total solidarity of systems or faster-
than-light influences between two separate systems—are they
truly two equivalent ways of expressing the same situation?
This is not certain, and the second seems in any case quite
imprudent, insofar as it may lead one to believe that the laws of
relativity would sometimes be transgressed. Indeed, everyone
recognises that experiments based on the 'paradox' of Einstein,
Podolsky, and Rosen manifest instantaneous *influences* at a distance, but in no case allow the transmission of *signals* contain-
ing information at a speed greater than that of light. This is due
to the strictly random nature of observations made in one and
the same place. For example, an observer placed near one of
the two analysers in the Aspect experiment never records anything but random responses, obeying the same law of probability regardless of the arrangement of the analyser examining the

8   B. D'Espagnat, "Use of Inequalities for the Experimental Test of a
     General Conception of the Foundation of Microphysics," *Physical
     Review D* 11 (1975), 1424; *Physical Review D* 18 (1977), 349.
9   B. D'Espagnat, "Les Implications conceptuelles de la physique
     quantique," *Journal de Physique* C2 (1981), 104.

polarisation state of the other photon in the other arm of the device. It is only by having the statistics concerning the measurements carried out on both photons at once that one can reveal the correlations characteristic of inseparability. Thus, the 'influence' between the orientation given to one of the analysers and the responses obtained to the other (and vice versa) does not constitute a 'signal'. It is therefore probably better to avoid, in a context that is known to be characterised by non-separability, speaking of the transmission of influences between 'distant' objects at a distance, when the notion of distancing presupposes that of distance, and therefore that of separability.

After Alain Aspect's pioneering experiments in France, the most recent experiments have brought to less than one chance in a million the probability of being mistaken in this conclusion and still being able to explain the observations while preserving the locality.[10] As soon as a composite system, formed by two quantum systems, has interacted, the two systems can no longer be considered as independent. Any measurement made on one of the two systems of a pair of quantities considered by Quantum Mechanics as not measurable together (such as position and velocity, or in a pair of photons the orientation of the polarisation in two different directions) immediately leads to a change in the quantum state of the second system, however remote it may be. According to Quantum Mechanics, each of the local measurements gives a result governed by a law of probability, apparently at random. But the correlations between the measurements made on each of the two entangled systems betray their solidarity, in a way that, as we have seen, cannot be explained by causality, whether one thinks of

---

10 L. Shalm et al.,"A Strong Loophole-Free Test of Local Realism," *arXiv: 1511.03189*, 2015. M. Guistina et al., "A Significant Loophole-Free Test of Bell's Theorem with Entangled Photons," *arXiv:1511.03190*, 2015.

a signal propagating (necessarily at a speed less than or equal to the speed of light) between the two measuring instruments or by a specific labeling of properties during the production of the entangled systems (the 'hidden variables' of Einstein and his collaborators).

On the other hand, the most recent and most demonstrative measurements all concern the production of pairs of entangled photons, whose polarisation is measured according to two variable orientations, with two measuring devices that can be separated by several kilometres. However, there are even more striking illustrations of this inseparability property, insofar as they do not concern photons, these evanescent light particles, but matter particles, whose entangled properties can be as concrete as their lifetime.[11] This is the case, for example, of couples of $K°$ and anti-$K°$ particles, which can be produced in many nuclear interactions such as a proton and an antiproton annihilation. The $K°$ and anti-$K°$ pair produced is entangled and can be described in quantum mechanics from the point of view of their lifetime as a superposition of $K°_S$ (s here means 'short living') and $K°_L$ (L for 'long living') particles; in fact, the second of these particles has an average lifetime five hundred times longer than the first. In a bubble chamber, one can thus (almost) always observe the disintegration of one of the particles at the end of a very short path, and the other at the end of a long path (an idealised sketch of such a picture is presented in Figure 7.2). But from the production point of view, each of these two particles K^0 and anti-K^0 always has at the origin a probability equal to one half of disintegrating according to one or the other mode. The question that arises is: once the short-lived particle has decayed, how does the other particle know that it must now

---

11 The lifetime of an unstable particle is the average time interval between its creation and its decay.

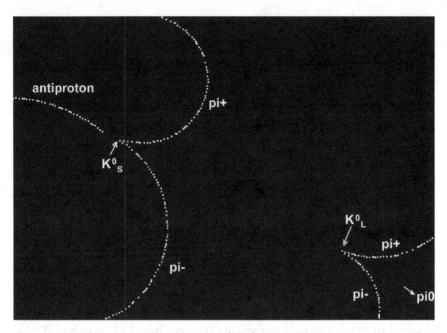

Figure 7.2. *Idealised sketch of a bubble chamber picture showing the annihilation of an antiproton entering from the left and annihilating with a proton in the chamber into two particles K^0 and anti-K^0, emitted at the point where the trace of the antiproton disappears. These particles, quantically entangled, do not cause the formation of bubbles in their passage, because they are not electrically charged. Each of them, however, is described in quantum mechanics as the superposition of two decay modes, one with an average lifetime of about 0.1 billionth of a second and thus a very short path (here, in a pair π^+ and π), the other with a lifetime 500 times longer (here, in three pieces π^+, π, and π^0). Quantically entangled, they must necessarily share the two disintegration modes. The intriguing question is 'once one of the two particles produced in the annihilation has decayed in the short mode, how does the other particle know that it must decay in the long mode?'*

disintegrate according to the other mode, into a long-lived particle? There is no solution to this dilemma other than to admit that the two particles form a single quantum 'object', regardless of their mutual distance.

One can object that these paradoxical observations always concern microscopic elementary particles. But non-locality has recently been verified for macroscopic systems, the only condition being that these macroscopic systems, carefully isolated from external influences, continue to obey quantum mechanics; that is, in this case, that their state is described in Quantum Mechanics by a superposition of states of two entangled, though locally separated, subsystems.

THE SOLIDARITY OF THE WORLD ACCORDING TO WHITEHEAD

Whitehead has always defended the solidarity of the world, understood as a global organism built by a converging and diverging tree of actual occasions that prehend each other. To be sure, he was not inspired for this, as far as we know, by the creators of quantum mechanics, beyond the 'discretisation' aspects of the theory (especially for the notion of 'atomic' time that we discussed in Chapter 4). But after his death in 1947, a number of philosophers and physicists noted the convergence between modern quantum theory and Whitehead's organic vision. Of particular note is the case of the physicist Henry Peirce Stapp, who worked directly with several of the founders of Quantum Theory, and then became very interested in Whitehead's philosophy.

Stapp was born in 1928 in Cleveland, USA; he obtained his PhD in particle physics at the University of California at Berkeley, under the supervision of two Nobel Prize winners, Emilio Segré and Owen Chamberley; both of them were awarded for the discovery of the antiproton. In 1958, he was invited to work with Wolfgang Pauli in Zurich and then in 1969 with Heisenberg in Munich. This meant that he frequented the most competent minds to discuss the foundations of quantum mechanics with them. When he discovered Whitehead's work in the 1970s, he was struck by the closeness of the philosopher's ideas to the conception of nature among the founders of quantum mechanics, particularly in Heisenberg.

In a 1984 article, Stapp explains:

> Quantum theory has, nevertheless, one feature that suggests that it should be formulated as theory of process. The wave function of the quantum theory is most naturally interpreted as representing "tendencies" or "potential" for actual events. This intuitive idea of the meaning of the wave function was first made explicit by David Bohm in his 1951 textbook, *Quantum Theory*. The idea was endorsed by Heisenberg and probably agrees with the intuitive ideas of most quantum physicists,

and a little further on Stapp adds:

> The process formulation of quantum mechanics contains no explicit dependence on an observer: it allows quantum theory to be regarded as a theory describing the actual unfolding or development of the universe itself, rather than merely a tool by which scientists can, under special conditions, form expectations about their observations.[12]

In a more recent paper, Stapp discusses a relativistic quantum ontology in close agreement with many of Whitehead's key ideas. 'Emphasizing these connections will flesh out the rational ontological construal of relativistic quantum field theory'.[13] Quantum field theory (Tomonaga, 1946) is indeed a natural extension of quantum mechanics to deal with the interactions

---

12 H. P. Stapp, "Einstein Time and Process Time," in *Physics and the Ultimate Significance of Time* (Albany: State University of New York, 1986), 264–67.

13 H. P. Stapp, "Whitehead, James, and the Ontology of Quantum Theory," *Mind & Matter* 5 (2007), 83–109.

between particles, their creation and their destruction, in a framework compatible with Einstein's special relativity.

In the same article, Stapp compares the explanation of the present in the Whiteheadian perspective with that proposed by quantum field theory (Figure 7.3).

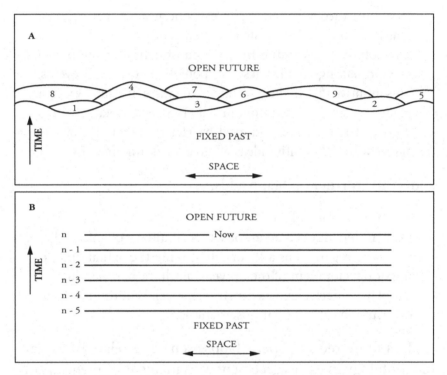

Figure 7.3. *Illustration comparing the interpretation of present tense constitution in Whitehead and in quantum field theory, based on Henry Stapp.*[14] *(A) Representation of the construction of the now in the Whiteheadian perspective. The wavy line at the base of this figure represents an initial 'present'. In the dynamic evolution of the quantum state this surface continuously increases towards the future, first of all by encompassing the spatio-temporal region marked 1. This unitary evolution, governed by the relativistic generalisation of*

14 Ibid., 90–91.

Figure 7.3. (*Continued*). *Schrödinger's equation, leaves unaltered the aspects of the initial quantum state. When a new concrescence — or, according to quantum mechanics, a new reduction of the wave function representing the portion 1 of space-time—occurs, it acts directly (as a projection) only on the new part of the surface represented by the upper limit of region 1. But this direct change also causes indirect changes on the present surface because of the prehensions related to this concrescence— or, in quantum terms, to quantum entanglements. These indirect changes produce what Stapp calls 'faster-than-light' effects which Einstein called 'spooky actions at a distance'. (B) This framework describes the corresponding treatment in quantum field theory. In both theories, the past is continuously enriched as it 'freezes', while the future remains open.*

Thus, insists Stapp, 'In the Whiteheadian ontologicalization of quantum theory, each quantum reduction event [of the wave function] is identified with a Whiteheadian actual entity. *In the quantum version*, such an actual entity performs two kinds of actions. An action of the first kind *partitions* a continuum into a collection of discrete *experientially distinct* possibilities. An action of the second kind selects (actualises) one of these discrete possibilities and obliterates the rest'.[15] Note that here, Stapp in fact succinctly describes the process of successive concrescences in Whitehead.

Henry Stapp adds to this comparison an interesting reflection, underlining affinities as well as oppositions of thought:

This conception of a growing actual space-time region— filled with (the standpoints of) the growing set of past actual entities—that advances into the potential open future constitutes a resolution to the famous debate between Newton and Leibniz about the nature of space. Newton's conception,

---

15 Ibid., 92–94.

described in the *Scholium* in his main work, "Principia Mathematica", was essentially a *receptacle* conception, in which space is an empty container into which movable physical objects can be placed. By contrast, Leibniz argued for a *relational* view that space is naught but relations among actually existing entities: Completely empty space is a nonsensical idea.

Didn't Whitehead say so himself? 'What I mean is that there are no spatial facts or temporal facts apart from physical nature, namely that space and time are merely ways of expressing certain truths about the relations between events'.[16]

Henri Stapp continues his paper by taking a certain distance, or at least by acknowledging the unfinished nature of Whitehead's philosophy when one wants to bring it closer to contemporary quantum theory; he is reticent about certain aspects of the theory of the organism, which are difficult to accept for a physicist who is used to the idea of separation between subject and object, between observer and thing. '[Whitehead's] theory is not implied by the currently available empirical data, but it gives a rationally coherent way to accommodate the discreteness aspects that Bohr and James identified. This, in spite of the fact that the theory specifies distinctive conditions pertaining to space, time, causation, the notion of the "now", the physically and psychologically described aspects of nature, and the nature of conscious agents. The empirically validated anthropocentric concepts of contemporary orthodox pragmatic quantum theory become thereby embedded in a general non-anthropocentric theory of reality'.[17] This is why, quite naturally, the physicist concludes his article by calling for

---

16 A. N. Whitehead, *The Concept of Nature* (Cambridge: Cambridge University Press 1920), 168.

17 Stapp, "Whitehead," 88–95.

a revised version of Whiteheadianism, which is still to come. It should be noted, however, that not all the reservations thus expressed by Stapp are shared by Whitehead's specialists.[18]

POTENTIALITY, REALITY, ACTUALITY

If Henry Stapp insisted so much on the similarities between Whitehead's philosophy and Heisenberg's thought, it is in particular because of the common distinction that both authors make between potentiality and reality. In his book *Physics and Philosophy*, Heisenberg thus writes:

> [...] in quantum theory all the classical concepts are, when applied to the atom, just as well and just as little defined as the "temperature of the atom"; they are correlated with statistical expectations; only in rare cases may the expectation become equivalent of certainty. Again, as in classical thermodynamics, it is difficult to call the expectation objective. One might perhaps call it an objective tendency or possibility, a "potentia" in the sense of Aristotelian philosophy. In fact, I believe that the language actually used by physicists when they speak about atomic events produces in their minds similar notions as the concept "potentia".[19]

---

18 I am thinking for instance of Dr. Joachim Klose, who responded to Stapp by writing: 'I believe that Henry Stapp did not sufficiently take into account Whitehead's discussion of prehensions when he claimed that Whitehead's system of metaphysics is incompatible with quantum theory due to Bell's theorem. In contrast, Bell's theorem could be used to support process philosophy.' J. Klose, 'Process Ontology from Whitehead to Quantum Physics', in *Recasting Reality. Wolfgang's Pauli's Philosophical Ideas and Contemporary Science*, ed. H. Atmanspacher and H. Primas (Berlin: Springer, 2009).

19 W. Heisenberg, *Physics and Philosophy, The Revolution in Modern Science* (New York: Harper & Brothers Publishers, 1958), 180.

It is therefore perhaps above all by their common rehabil-
itation of Aristotle's 'potentia', their common insistence on the
notion of potentiality and their agreement to give it a degree
of reality (but not of actuality) that Heisenberg and Whitehead
resemble each other. Both gave the wave function a value
of 'potentiality'; their difference, notes Stapp, is that 'when
Heisenberg gives an active role to the observing agent', so that
the transition from possibility to actuality takes place in the act
of observation, 'for Whitehead concreteness is an objective fact,
independent of the observer'. This is an opportunity for us to
underline once again the fundamental importance accorded to
potentiality in Whitehead, who defines his *ontological principle*
with these words: 'it is the principle that everything is positively
somewhere, and in potency everywhere'.[20] For him, everything
that is upstream of concrescence—actual entities of the past
(and therefore 'dead', but still real according to Whitehead),
'eternal' entities and their possible mutual prehensions with
entities in the process of concrescence—all this is, of the order
of potentiality, has no place or time. Only is concrete what hap-
pens in the very act of concrescence, the fleeting crystallisation
of the real, a concrete act of nature itself and not of the mind.
Philosophers and theoreticians of quantum mechanics are still
divided on this point, which led to the famous 'Schrödinger's
cat' controversy; but to enter here into this debate would take
us too far.[21]

Closer to home, the similarities in thinking between
Whitehead and Heisenberg noted by Henry Stapp are also noted

---

20 A. N. Whitehead, *Process and Reality* (New York: Macmillan,
     1929), 40.
21 The interested reader may refer to the state of play of this controversy,
     given in the article "Schrödinger's Cat" in the online Encyclopedia
     Wikipedia.

by Marc Lacoste de Lareymondie and Sébastien Poinat. The first, a physicist and author of a thesis on Whitehead, comments:

> We do not know if Heisenberg had been aware of Whitehead's work, which he does not quote. But we do recognize the major themes of Whitehead's cosmology: criticism of the notion of substance and its corollary, the subject-object distinction; a return to pre-Kantian ways of thinking; refusal of a dogmatic position and a permanent concern to take into account the progress of scientific and philosophical knowledge without neglecting 'the relevance to the stubborn facts of daily life.[22]

For his part, the philosopher Sébastien Poinat, in his article "Whitehead et les pères fondateurs de la Mécanique Quantique"[23] also notes the central role of the distinction between the potential world and the present world in the philosophy of these two thinkers:

> Schrödinger's equation and all quantum mechanics can be seen as the description of the temporal evolution of the real potentialities that make up reality. Quantum mechanics would thus not so much deal with the actualized reality, that of the now, as with these powers of phenomenal manifestation. Heisenberg's reading would thus lead us to consider that quantum mechanics accomplishes the same gesture as Whitehead's philosophy: understanding actuality from potentiality.[24]

---

22 M. Lacoste de Lareymondie, *Une Philosophie pour la Physique Quantique. Essai sur la non-séparabilité et la cosmologie de A.N. Whitehead* (Paris: L'Harmattan, 2006), 311.

23 S. Poinat, "Whitehead et les pères fondateurs de la Mécanique Quantique," *Noesis* 13 (2008), 175–91.

24 Ibid., 184.

For Whitehead, both worlds are real, but only the second, resulting from a global concrescence, is 'concrete' or 'actual'. And Poinat quotes a 1942 reflection by Heisenberg: 'Quantum theory is that idealization where reality appears at every instant as a determined abundance of possibilities for an objective actualization'.[25] From there, the author concludes that 'Being is therefore fundamentally understood from the notion of potentiality and actuality is nothing more than the passage between past potentialities and future potentialities'.

All in all, we are obliged to note that Whitehead, although never having collaborated directly in the great work of building quantum mechanics, but certainly inspired in part by it, has given the world a philosophy of nature that can help us understand many seemingly paradoxical aspects of quantum mechanics. Whitehead himself explicitly aimed at this goal. Indeed, he argues in *Process and Reality* that his theory of the organism provides a conceptual framework for quantum theory.[26] In his theory, it is the prehensions that ensure the solidarity of the world and prevent future actual occasions to have a precise localisation before their concretisation.[27] However, it must be recognised that the total reconciliation of Whitehead's philosophy of the organism with quantum mechanics faces a major obstacle, which has been well emphasised by the philosopher of science and physicist Abner Shimony. For Whitehead's philosophy totally ignores one of the key principles of wave mechanics, the basis of the understanding of quantum entanglement phenomena: the principle of superposition of states.

---

25 W. Heisenberg, *Philosophie. Le Manuscrit de 1942* (Paris: Le Seuil, 1998), 310.
26 Whitehead, *Process and Reality*, 78–79, 117.
27 Ibid., 117.

The Whiteheadian treatment of the state of a composite system is at odds with a quantum mechanical principle which has attracted little attention in spite of its revolutionary philosophical implications: *that a several-particle system may be in a definite state, i.e. may have as definite properties as quantum theory permits, without the individual particles [that compose it] being in defined states.*[28]

In this article, Shimony speculates on the idea that Whitehead's philosophy could perhaps help us understand, better than the current treatment of measurement theory (decoherence theory), the mechanism by which the reduction of a superposition of states occurs, adding that this would probably require replacing the fundamental equation of quantum mechanics, the Schrödinger equation, by a stochastic generalisation of this equation (p. 261). A gigantic work that has not yet been undertaken. But perhaps the challenge is worth it, if it led to a new general theory of nature, precise and quantitative, of Whiteheadian inspiration, and rid of the still obvious contradictions between the two current physical theories: quantum mechanics and relativity.

---

28 A. Shimony, "Quantum Physics and the Philosophy of Whitehead," in *Philosophy in America*, ed. Max Black (Ithaca: Cornell University Press, 1965), 251.

CHAPTER 8

The Space–Time of Relativity and Whitehead's Extensive Continuum

SPECIAL AND GENERAL RELATIVITY: FROM EINSTEIN TO WHITEHEAD[1]

Special Relativity

Let us recall that one of the principles of classical mechanics, due to Galileo, concerns the addition of speeds: if a bow on land projects its arrow at speed V and the arrow is shot forward from a boat moving at speed v, the speed of the arrow in relation to the land will be V + v. On the other hand, Maxwell's equations describing the propagation of an electromagnetic wave (e.g., light) have only one value for its propagation speed, the value c, whatever the conditions of emission of this wave and in

---

1   The present account of the elaboration of the theories of special and general relativity is here largely borrowed from my book *Les Fils du Temps* (Paris, France: CNRS Editions, 1990, 2016)—*The Children of Time* (University of Illinois Press, 1995).

particular the speed of its source. When Einstein reflected on this contradiction in 1905, he realised that it could be resolved if the notion of time bequeathed by Newton was radically challenged. By redefining time in its relations with space, Einstein had indeed realised that the new theory could preserve in their exact form Maxwell's equations, setting the laws of electromagnetism and providing for the constancy of the speed of light in all inertial reference frames. He then considered this property as an intangible result and posed the following two postulates as fundamental principles of any future physical theory:

- *Postulate of relativity*: The Newtonian notion of absolute rest, that is, rest in relation to absolute space, is unfounded because it corresponds to nothing observable. Electrodynamic or mechanical phenomena do not have any properties corresponding to this idea. The notion of 'ether', required to materialise this absolute rest in the case of electromagnetic waves, is superfluous and can therefore be abandoned.
- *Postulate of the constancy of the speed of light*: The speed of light is independent of the motion of its source, as explicitly stated in Maxwell's theory.

Once again, in the framework of Newtonian absolute time, these two postulates would be incompatible with each other. Indeed, the Galilean law of velocity composition would state that if the speed of the source measured in the system with the ether at rest (system in absolute rest) is v, then the speed of the emitted light, measured in this same frame of reference, must be c + v. This rule, used implicitly, would lead to the curious results predicted by Einstein as a teenager when he imagined himself looking in a mirror in a very fast vessel. Reconciling the two previous postulates therefore requires a revision of the basic principles of mechanics. Revision is possible insofar as the choice set by Newton for the location of events in space and time is arbitrary and can be appropriately modified. In fact, the two postulates precisely and uniquely impose the form that this modification must take: the one codified in the 'Lorentz transformations', which specify how the space and time coordinates of an event observed from two reference frames moving rectilinearly and uniformly with respect to each other change.

The postulate of relativity itself, to which Einstein's theory owes its name, is the first of the two postulates stated above. It is in fact a natural extension of the postulate of Galilean relativity, according to which one cannot, by any mechanical experiment, detect the motion of the laboratory in which one carries out this experiment if this motion is rectilinear and uniform with respect to 'absolute space' (usually materialised by a reference system linked to 'fixed stars'). Einstein's postulate of relativity extends the above statement to all physics experiments, whether mechanical or electromagnetic in nature.

In his June 1905 article,[2] the statements of the young patent inspector of Berne are written in a form that seems to limit their epistemological scope. He challenges the notion of absolute rest, but does not yet state that 'the form of the equations of physics must remain the same in all the reference frames of inertia'. He asserts the constancy of the speed of light, but makes no reference to a limiting velocity of energy transfer in the vacuum or of information transfer. He limited himself to a technical presentation of the results obtained on the basis of the two previous postulates, in particular, in the context of

---

2   A. Einstein, "Zur Elektrodynamik bewegter Körper," *Annalen der Physik* 17(1905), 891–921. English version available on https://www.fourmilab.ch/etexts/einstein/specrel/specrel.pdf

the electrodynamics of moving bodies. Thus, in 1905, Einstein does not yet seem to have fully perceived the true nature of the revolution he has just introduced. The criticism of the notion of simultaneity is based on the propagation of light signals, and the postulate of relativity places the propagation of light at the centre of physical theory. But why make the speed of light play such a role in the description of nature? What is the origin of this alleged despotism, reinforced *a posteriori* by the stated result that the speed of light is a limiting speed that no material body can reach, because it would take infinite energy to push it that fast? Would one say that almost all of our experiences are mediated by sight and the use of light rays? Insufficient explanation: the physics that a population of blind people would develop would undoubtedly be formally the same as ours. We therefore need another foundation for the pre-eminence of the speed of light.

Today, most epistemologists believe that Einstein actually introduced in 1905 a physical theory whose touchstone is the paradigm of causality propagated at velocity c. In the theory of relativity, time is a parameter that allows us to label in increasing order the points-events joined by a causal chain or which can in principle be joined by a causal chain. The versatility of the different ways of charting in time and space in accordance with the Lorentz transformations is the consequence of this absolute order regulated by the propagated causality. The theory of relativity is therefore a *causal theory of time*.

In fact, it is not Einstein himself who took the first steps in this direction, but Hermann Minkowski, his former professor. In 1907, he published a paper entitled "The Fundamental Equations for Electromagnetic Processes in Moving Bodies" and the following year, he gave a resounding lecture on 'Time and space'. In it, he launched the provocative apostrophes that we recalled in Chapter 4: 'Gentlemen! The concepts about time and space, which I would like to develop before you today, have grown on experimental physical grounds [...] Henceforth, space for itself, and time for itself shall completely reduce to a mere shadow, and only some sort of union of the two shall preserve independence'.[3]

After formally introducing four-dimensional space–time (where the fourth dimension is time converted into space, thanks to the conversion constant c), Minkowski divides it into zones called 'past', 'future', and 'elsewhere'. Only events in the past are perceived by an observer located at the origin, and the latter can only influence events located in the future. The events taking place elsewhere can neither be known to him nor influenced by him (Figure 8.1). On the other hand, two events located in the elsewhere of each other can be declared successive by one observer and following each other in the opposite order by another. Science-fiction authors have hastily concluded that 'time is relative'. This is of course not the case, there is no causal influence between these two events, no matter who the observer is. And for a couple of events of the cone of light, the order of succession is immutable for any observer whatever his speed of movement.

Minkowski showed next that, while spatial and temporal measurements vary when moving from one inertial coordinate system to another, one quantity defined as the 'invariant

---

3   H. Minkowski, English translation of the talk "Space and Time" given in 1908 in Cologne, 1909, https://en.wikisource.org/wiki/ Translation:Space_and_Time. The quoted sentence may rightly seem excessive. The notion of proper time marking the evolution of a system, observed in a reference point in relation to which this system is at rest, is enough to mark the specificity of time in relation to space.

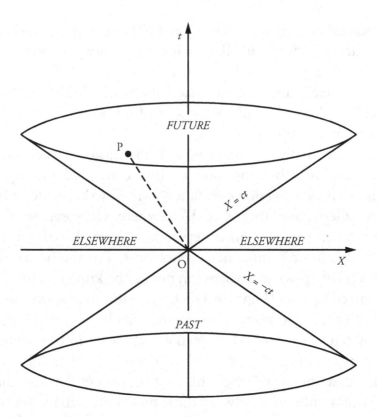

Figure 8.1. *Minkowski's relativistic space–time. Time has been represented on the ordinate and the x-coordinate on the abscissa (y and z coordinates are not represented). The observer is assumed to be located at the origin and at time t = 0. For him, the separation of the relativistic space–time in past, future, and elsewhere is delimited by the 'light cone' whose trace on the represented plane is given by x = ± ct. The separation between the point of origin and the event P of coordinates (x, y, z, t) is measured by their 'invariant interval'. For the points located on the light cone, this invariant interval is zero. As customary, the light cone has also been represented in perspective in three-dimensional projection (x, y, t).*

interval'[4] remains constant despite the change of reference frame. It is therefore a constant for all inertial reference frames. The invariant interval between two event-points located on the trajectory of the same ray of light is zero. For an observer, event-points located in the past or in the future are at a negative (squared) invariant interval. Those of the 'elsewhere' are located at a positive invariant interval. Thus, in relativity theory, the use of light rays, as the essential carriers of causal influences, determines the structure of the universe.

General Relativity

The Relativity introduced in 1905 by Einstein is qualified as 'special' because it applies only to transformations between two inertial reference frames, one pointing to fixed stars and the other moving at constant speed with respect to the first. Reflecting on a possible extension of his theory, Einstein observed in 1907 that by establishing as a universal principle the equality of weight masses (the reciprocal attraction of masses) and inertial masses (their resistance to acceleration), he could build a general theory of Relativity, applicable in any coordinate system, even if it undergoes an acceleration or a rotation with respect to an inertial reference frame. Thus, the equality of inertial mass and weight mass, far from being a fortuitous circumstance, assumes for him the value of a principle: that of the relativity of gravitation. It allows him to interpret gravitational attraction as a geometric effect. Of course, this geometry must be generalised, it must include in the definition of distances between bodies, an element that reflects the presence of sources of the gravitational field.

---

4   The invariant interval between two event-points is given by the formula $d^2 = x^2 + y^2 + z^2 - c^2 t^2$. This definition generalises, in a four-dimensional space–time, the notion of spatial distance in Cartesian three-dimensional space.

The concrete elaboration of this theory, in which the geometry is determined by the sources of the gravitational field, proved to be laborious. Marcel Grossmann, an old friend of Einstein, had mastered the techniques of calculating non-Euclidean geometries. Both collaborated to apply them to the new relativistic theory of gravitation that Einstein was trying to build.[5] In four years (1911–1915), Einstein published the first articles on the theory of general relativity: the principle of equivalence was discussed in detail again in 1911, the notations of Riemannian geometry were introduced in an article written in collaboration with Grossmann in 1913, and the general equations of gravitation were presented to the Prussian Academy of Sciences in 1915. These equations allow the writing of the laws of physics both in a uniformly accelerated frame of reference and in any frame of reference: they are in accordance with Einstein's principles, according to which 'the choice of a frame of reference must be indifferent for the expression of the laws of Physics'.

Practical Consequences on the Behaviour of Clocks

In the theory of general relativity, time keeps strictly the same status as in special relativity: it is a causal parameter related to the propagation of light. Moreover, in the immediate vicinity of an event-point, one can always give oneself a reference frame such as the metric to be applied—the prescription for the measurement of distances, taking into account the

---

5   Einstein assimilated the techniques in question rapidly and well but without feeling an immoderate love for the mathematics per se. He was no doubt thinking about this time in his life when, responding in jest to a little girl who had told him about her troubles with mathematics, he said 'Do not worry about your difficulties in mathematics; I can assure you that mine are still greater!'

sources of gravitation present—would be that of special relativity: in particular, light propagates in a straight line with the speed c.

Over large distances and in the presence of massive objects, however, light follows a trajectory that does not correspond to a straight line in Euclidean space. Thus, the rays emanating from an object located behind the sun and grazing the edge of its disk undergo a deviation whose magnitude, predicted in 1913, was verified in 1919 during a total solar eclipse. Like space intervals, time intervals lose the property of congruence (superimposability in a displacement) characteristic of Euclidean geometry. In the vicinity of massive bodies, or during accelerations, clocks slow down.

In 1915, Einstein is thus in possession of the general principles and fundamental equations of the relativistic theory of gravitation. Like special relativity, this theory rejects the existence of an absolute reference frame for space and time (Euclidean absolute space and Newton's absolute time). It further asserts that metrics (the way of calculating distances in space–time) can be chosen at will so that the laws of physics can be expressed as simply as possible, as long as this choice is confirmed by experiment. In general relativity, the trajectory of bodies free of any force other than universal gravitation is represented by a 'geodesic', that is, the natural extension of the straight lines of Euclidean geometry. The originality of the theory is to eliminate, amongst the forces of nature, the gravitational attraction, which thus becomes a purely geometrical effect. In other words, for Einstein, geometry is only the expression of the relationships between massive objects or sources of energy in the world; geometry is the theory of that. This is an epistemological point of view that Whitehead will challenge. Ronald Desmet explains:

In Einstein's interpretation of the physics of relativity the bifurcation of classical physics is reconfirmed: nature is split in two worlds; on the one hand, the world of so-called objective science, a postulate of our theoretical thought; and on the other hand, the world of subjective perception and common sense, the basis of our praxis. Whitehead has fought this bifurcation, and he replaced Einstein's interpretation with one that is coherent with the presupposition of common sense that the geometry of space-time is uniform, and independent of the physics of gravitation.[6]

At the time of its publication, Einstein's new theory aroused little curiosity in the scientific community, apart from a small circle of physicists. Yet it already explains some remarkable effects, such as the advance of the perihelion of the planet Mercury (an effect already known at the time, but which remained unexplained in the framework of Newton's theory), and predicts the curvature of light rays in the vicinity of the sun. The verification of this effect in 1919 marks the real beginning of the notoriety of its author. On the other hand, at that time the delay of clocks in a gravitational field or during accelerations was not yet verified.

Whitehead Reads of Einstein's Work

As Ronald Desmet tells us,[7] Whitehead became familiar with the 1905 theory of special relativity only around 1912–1914, as did most of his British contemporaries. His first reference to the theory of special relativity seems to be

---

6   R. Desmet, "Whitehead's Principle of Relativity," in *Whitehead—The Algebra of Metaphysics*, ed. R. Desmet and M. Weber (Bruxelles: Chromatika, 2010), 210.

7   Ibid.

a lecture entitled "Space, Time and Relativity" that he gave to the British Society for the Advancement of Science in 1915. He became acquainted with the theory of General Relativity in 1916, probably through lectures by Arthur Eddington and Ebenezer Cunningham. He himself recounted how he became aware of the first confirmation of the deviation of light rays by massive bodies: 'I was present at the meeting of the Royal Society in London when the Royal Astronomer for England announced that the photographic plates of the eclipse of 1919, measured by his colleagues at the Greenwich Observatory, had verified Einstein's prediction of the curvature of light rays in the vicinity of the sun'.[8] He wrote *The Principle of Relativity with Applications to Physical Science* in 1922 with the aim of reformulating Einstein's theory of gravitation in such a way that gravitation is no longer identified with the proposed curvature of space–time, but as a physical interaction in the traditional sense, within Minkowski's uniform space–time framework. In early June 1921, Whitehead had the opportunity to discuss his proposal with Einstein himself, when the latter made a brief visit to one of Whitehead's best friends in London, Richard Haldane. Unfortunately, the content of their conversation remains unknown to us. But Whitehead had many opportunities afterwards to express his reservations and to propose an alternative to Einstein's theory.

WHITEHEAD CRITICISES EINSTEIN'S APPROACH TO RELATIVITY

It is clear that Whitehead thought that the latter did not go far enough in his criticism of the Newtonian notion of time.

---

8   A. N. Whitehead, *Science and the Modern World* (New York: Macmillan, 1925), 13.

Einstein's Theory of Relativity accounts for the causal structure of the universe, once the invariability of the speed of light, the carrier of causal actions, has been accepted. But Whitehead, while admitting this fact, wanted to base the theory more firmly on common and immediate experience. While he recognised that Einstein's theory of Relativity represented 'an opportunity to transform classical physics into a new science: the science of the global unity of natural entities', he did not like Einstein's way of introducing the curvature of space–time to account for the gravitational effects of mass and energy. In *The Concept of Nature*, he writes:

> The divergence [between us] chiefly arises from the fact that I do not accept his theory of non-uniform space or his assumption as to the peculiar fundamental character of light-signals. I would not however be misunderstood to be lacking in appreciation of the value of his recent work on general relativity which has the high merit of first disclosing the way in which mathematical physics should proceed in the light of the principle of relativity. But in my judgment, he has cramped the development of his brilliant mathematical method in the narrow bounds of a very doubtful philosophy.[9]

For Whitehead, it is indeed excluded that objects can causally influence space–time; for him, uniformity is essential for knowledge and seems a necessary condition so that measurement be possible (to ensure the congruence of rulers and clocks—a property lost in general relativity, as we have seen).[10] Thus, the Whiteheadian extensive continuum is necessarily of

---

9   A. N. Whitehead, *The Concept of Nature* (Cambridge: Cambridge University Press 1920), vii.

10   See on this respect J. Bain, "Whitehead's Theory of Gravity," *Studies in History and Philosophy of Modern Physics* 29 (1998): 547–74.

the Minkowskian type, involving its metric with constant coefficients. In his book on Relativity published in 1922, Whitehead explained well that if one wants to describe nature in a systematic way, one must necessarily adopt a metric (a way of calculating distances in space–time) that is uniform throughout space.[11] He gives two arguments in support of this position. On the one hand, indeed, he links the question of uniformity to the problem of measurement, through the obligatory notion of congruence: 'for it must remembered that measurement is essentially the comparison of operations which are performed under the same set of assigned conditions. If there is no possibility of assigned conditions applicable to different circumstances, there can be no measurement.'[12] On the other hand, space and time are abstractions from the extensive relations between events. The structure of these necessary relations must be uniform so that the knowledge of any part of the structure makes it possible to know the structure in its totality: this condition seems to him indispensable for induction to be possible. He further developed this argument in *Science and the Modern World*, published in 1925.

That's not all; there is more between Einstein and Whitehead. From the beginning, in fact, the basic concepts of the two thinkers diverge. In *The Concept of Nature*, Whitehead devotes lengthy developments to give status to the notion of simultaneity. Far from rejecting this concept, he sees it as a general fact that is binding on any observer (although he carefully distinguishes it from the concept of instantaneity). Let us recall that in this work he wrote: 'simultaneity must not be conceived as an irrelevant mental concept imposed upon nature. [...]

---

11 A. N. Whitehead, *The Principle of Relativity with Applications to Physical Science* (Cambridge: Cambridge University Press, 1922), 73.
12 A. N. Whitehead, *The Interpretation of Science, Selected Essays* (Indianapolis: Bobbs-Merrill Co., 1961), 134.

A duration is a concrete slab of nature limited by simultaneity which is an essential factor disclosed in sense-awareness'.[13] Constrained by the Einsteinian critique of simultaneity, which emphasises that this concept is not operationally applicable, Whitehead abandoned the name simultaneity to replace it with contemporaneity, which naturally implies a slice of time that is all the wider the broader the scope of the experience.

The fundamental equation of General Relativity, which connects the metric, the invariant distances, and the curvature of space–time at all points to the gravitational field, and thus to matter and energy density in all space, seemed to him to contain a vicious circle like a snake biting its tail. Indeed, 'one might say Einstein supposes that matter (or the gravitational field) is ontologically prior to space–time, [...] while events are simply intersections of world-lines of particles; and Whitehead supposes that events are ontologically prior to space–time, while matter is merely a contingent characteristic of certain events'.[14]

DEVELOPMENT OF A RIVAL THEORY OF GRAVITATION

How did Whitehead—who received at Cambridge a very strong formation in mathematics under the guidance of William Niven and Edward Routh—meet the formidable challenge of developing a theory of gravitation competing with Einstein's general theory of relativity? First, through an intimate knowledge of Maxwell's electromagnetism, and then about the techniques related to the treatment of delayed potentials and the

---

13 Whitehead, *The Concept of Nature*, 53.
14 R. Palter, *Whitehead's Philosophy of Science* (Chicago: The University of Chicago Press, 1960), 213.

principle of least action, two important chapters of the physics of the propagated phenomena.

Thus, as early as 1919, Whitehead derived Lorentz's transformation laws from very general considerations of direct experience considered within the framework of his own metaphysical ontology. For him, explains Bain in the article quoted above, the meaning of space and time does not derive from theory, they are primary: 'What we mean are physical facts expressible in terms of immediate perceptions; and it is incumbent on us to produce the perceptions of those facts as the meanings of our terms'.[15]

The Extensive Space–Time Continuum

Whitehead does not give the word 'relativity' the same meaning as Einstein. In the language of the former, the main meaning of this word is 'relation' and the theory of relativity is above all a theory of the interrelation between events. The geometry of space–time is an expression of the uniform inter-relation that binds together all events in a single world, it is a consequence of that 'philosophy of the organism' which he holds so dear. For the latter, relativity is above all the principle that the laws of nature must not depend on the reference frame used to describe it, they are an intrinsic description of the structure of the universe. For him, geometry is a free construction of the mind. It is intended to express a hidden structure of the fabric of the universe, that is, relations between events. He first discovered that by revising the Newtonian concept of time, he could describe space–time as the framework of the universe, subject to the law of propagated causality. Then he discovered

---

15 A. N. Whitehead, *An Enquiry Concerning the Principles of Natural Knowledge* (Cambridge: Cambridge University Press, 1919), 46.

that this geometry could also account for universal gravitation, provided he abandoned the pre-supposition of Euclidean or pseudo-Euclidean space–time geometry. In a sense, Whitehead is much more conservative, because he insists on keeping the pseudo-Euclidean space–time suggested by primitive experience, corrected by that of causality. So what is the status of Whitehead's extensive continuum? In *Physics and the Ultimate Significance of Time*, the philosopher Patrick Hurley of the University of San Diego in the United States attempts an explanation: for Whitehead and contrary to Einstein, the extensive continuum does not represent the structure of the current world, it exists only to the extent that events exist. Thus:

> Prior to its actualization through ingression in actual occasions, the extensive continuum is only potential. This means that, prior to actualization, the potential extensive continuum is either an eternal object or a collection of eternal objects. As such, it is not even extensive—just as the concept of triangularity is not itself triangular. When the extensive continuum is actualized, extensiveness makes its appearance.[16]

And, once the concreteness is achieved and satisfaction is attained, as in the story of Thom Thumb, a white pebble, a four-dimensional brick appears in the objectified world in an immediate presentation. As far as time is concerned, this pebble expresses a fundamental trait of the Whiteheadian theory of time, which, as we have seen, excludes the instant and admits as actual only discrete durations.[17]

---

16 P. Hurley, "Time in the Earlier and Later Whitehead," in *Physics and the Ultimate Significance of Time* (D. R. Griffin ed., Albany, State University of New York, 1986), 102.

17 This white pebble or elementary brick is not a 'quantum' in the four-dimensional space–time because its volume depends on the considered concrescence.

As for gravitation, Whitehead in his theory describes it as a delayed remote action, not directly related to the geometry of space–time. Consequently, the system of equations he developed for his theory does not resemble at all Einstein's equations. The fundamental equation of the Whiteheadian theory of gravitation at a material point of space–time comprises two terms, the first of which accounts for the structure of relativistic geometry (Minkowski's pseudo-Euclidean space–time) and the second expresses the propagated influence of all the masses of the universe on the test mass at this point, which makes it possible, step by step, to deduce its trajectory (its world line).

It is remarkable that Whitehead's theory gives identical results to those of Einstein in simple cases, as Eddington showed two years after the publication of *The Principle of Relativity*.[18] More precisely, Whitehead's theory gives the same result as Einstein's General Relativity in the two sensitive tests that were within the reach of the measuring instruments of the time: a correct value of the precession of the orbit of the planet Mercury and a correct measurement of the deviation of the light rays grazing the surface of the sun. So far, so good for Whitehead. But, more recently, things have become more complicated for his theory.

FAILURES OF THE WHITEHEADIAN THEORY AND THEIR EPISTEMOLOGICAL MEANINGS

For today, it must be said that the Whiteheadian theory of relativity does not pass the test of experimental measurements made accessible by the increase of our knowledge about the universe and the current precision of our instruments. A certain

---

18 A. S. Eddington, "A Comparison of Whitehead's and Einstein's Formulae," *Nature* 113 (1924): 92.

number of these measurements have not given the predicted results, and are on the contrary in agreement with Einstein's theory of General Relativity.[19]

Three tests are particularly blatant. First, Whitehead's theory predicts an anisotropy of the 'effective' constant of gravitation in the galaxy, which would be due to the disk shape of the galaxy, and the associated the anisotropy of the distribution of matter within it. Second, it predicts an acceleration of the moon on its orbit, related to the difference in mass between the earth and its satellite, which should progressively move the satellite away from the earth. But the very precise distance measurements made possible by the installation of laser reflectors on the lunar ground largely disprove the expected effect. Last but not least, the theory predicts an acceleration of the centre of mass in a binary system, such as two stars or twin pulsars, an effect recently denied in the case of pulsar B1913+16 'by a factor of one million'.

At the present time, the devil does not seem to be in the details, but rather points some flaw in the basic philosophy. It is clear, indeed, that in Whitehead's philosophy, the laws of nature depend on the complex interactions between actual entities in a changing world. Consequently, the so-called fundamental constants of physics (the speed of light, the charge of the electron, the gravitational constant, the Planck constant of Quantum Mechanics, the fine structure constant α, etc.) in all likelihood should have varied over the course of the broad evolutions that the universe has undergone, its expansion in particular. But several of these constants have been (indirectly) measured and do not seem to have varied for billions of years,

---

19 G. Gibbons, and C. Wills, "On the Multiple Deaths of Whitehead's Theory of Gravity," *Studies in History and Philosophy of Modern Physics* 39 (2008): 41–61.

almost since the beginning. This is what many experts, such as Jean-Philippe Uzan and Bénédicte Leclerc, conclude in their book *De l'Importance d'être une Constante.*[20] Thus, one is entitled to conclude that the philosophy of nature presented by Whitehead does not agree with nature itself, at least in some of its aspects. Doesn't this remind us of the misunderstanding that arose between Einstein and Bergson when they met in Paris in 1922, the very same year as Whitehead's publication?[21] On the other hand, the fact that the Whiteheadian theory of gravitation has been 'falsified' (in the Popper's meaning of the term) does not prove that all its metaphysical pre-suppositions, and in particular its metaphysics of Time, are no longer worth considering and developing, at the risk of having to correct some of them. It should be noted, however, that the equation proposed by Whitehead is not the only possible theory of gravitation that accounts for delayed (propagated) gravitational actions within the framework of Whitehead's epistemology. Whitehead himself considered other possibilities, but preferred the one we are discussing here. The development of an alternative theory of relativity and gravitation based on largely Whiteheadian premises remains a task to be accomplished, next to the one we have mentioned in connection with its necessary accommodation with quantum mechanics.

---

20 J.-P. Uzan and B. Leclercq, *De l'Importance d'être une Constante* (Paris, France: Dunod, 2005), *The Natural Laws of the Universe: Understanding Fundamental Constants* (New York: Springer Publishing Company, 2008).
21 Read on that subject E. During, *Bergson et Einstein, la querelle du temps* (Paris, France: PUF, 2011).

CHAPTER 9

Process and Time

THE ENGINE OF THE WORLD IS NOT TIME, BUT THE PROCESS

As we have explained at length, time (be it our time or the mathematical time of physicists) is an abstraction. It cannot therefore be considered as one of the basic elements of reality, or even of concreteness. Is it therefore necessary to free ourselves from time when we think about reality? That would be an absurdity. Whitehead invites us to share this feeling, when he writes: 'It is nonsense to conceive of nature as a static fact, even for an instant devoid of duration. There is no nature apart from transition'. Yes, but this is to add immediately, 'and there is no transition apart from temporal duration'.[1] Time is therefore something of a transition, but it differs from it, a bit like Aristotle's thought, who wrote 'time is the number of motion'.[2] Naturally, Whitehead criticises classical science for basing its development on the Newtonian notion of time, because, as he wrote: 'I cannot in my own knowledge find anything corresponding to the bare time of the absolute theory. Time is

---

1   A. N. Whitehead, *Modes of Thought* (New York: Macmillan, 1938), 207.
2   Aristotle, *Physics*, Book IV (tr. R. P. Hardie and R. K. Gaye, accessed November 23, 2021, http://classics.mit.edu/Aristotle/physics.4.iv.html).

known to me as an abstraction from the passage of events. The fundamental fact which renders this abstraction possible is the passing of nature, its development, its creative advance'.[3] In writing *The Concept of Nature*, he decided to call this flow of nature *process*, and immediately committed himself to always distinguish between process and time.

The process is the means by which what is potential becomes an actual reality. It is the process by which the current particle-events or actual entities we have been talking about so far manifest both permanent characters and, at times, sparks of novelty. The process is the permanent river that carries us and carries us away, and at the same time the touchstone that reveals the concrete reality. Whitehead wants this new concept, contrary to that of time, to suit his concern to always start from concrete reality as it offers itself to our senses, and to allow, even better than the materialistic description of the physics of his time, to discover and enunciate the laws of nature.

Thus, it is the actual entities, uninterruptedly sawn apart or produced in disorder by this cosmic fountain that is the process itself, that create time. Each new entity adds by being born by concrescences some dust of time to the counter, because there are no instants. This is also what Maria Teixeira says, adding the idea of an always unfinished nature, always open to the future, in short, of temporality: 'The actual entities thus constitute temporality itself, and for this reason, the nature of the latter can only be unfinished. Time does not exist before actual occasions come into existence; it is the process that creates time itself. Temporality coincides with the actual occasions that come into existence and whose incompleteness pertains to

---

3   A. N. Whitehead, *The Concept of Nature* (Cambridge: Cambridge University Press, 1920), 17.

their nature, because they create time and time is intrinsically an open process.'[4]

I wrote that each concrescence adds some time dust to the counter. I deliberately avoided using the word clock, this too familiar instrument, too attached to our daily life. For it is now time to put our description in a four-dimensional framework. Of course, concrescences not only create time, they also create space, and each concrescence actually adds both a piece of time and a piece of space to the space-time continuum, as is clearly expressed in Figure 7.3, which we borrowed from Henry Stapp. Or more precisely, the process, by causing these cascades of actual entities in their four-dimensional extension, allows us to abstract from the formed reality this four-dimensional frame-work that we call space-time. Whitehead prefers to qualify it as an extensive continuum, in which the separation in space and time is relative, as Albert Einstein showed us. In *Process and Reality*, Whitehead warns:

> There is a prevalent misconception that 'becoming' involves the notion of a unique seriality for its advance into novelty. This is the classic notion of "time", which philosophy took over from common sense. Mankind made an unfortunate gener-alization from its experience of enduring objects. Recently physical science has abandoned this notion. Accordingly we should now purge cosmology of a point a view it ought never

---

4   M. T. Teixeira, *Ser, Devir e Perecer. A Criatividade na Filosofia de Whitehead* (Lisbon: Centro de Filosofia da Universidade de Lisboa, 2011), 97. Here we encounter a double idea: (*a*) we must not forget that fundamental events are always abstract sets, whose spatial and temporal extension depends on the degree of miniaturisation to which it is led; and (*b*) no concrescence gives rise to complete satisfaction. The thirst for harmony that permeates the world is never satisfied. Each concrescence therefore calls for another, which will replace it.

to have adopted as an ultimate metaphysical principle. In these lectures the term "creative advance" is not to be construed in the sense of a uniquely serial advance.[5]

Does he have in mind creative advances that would concomitantly concern actual occasions located in different regions of the extensive continuum? Certainly, a world of current opportunities that would 'prehend' each other mutually without creative advances, without the appearance of novelty, would be of no interest, and this alone justifies Bergson's aphorism 'time is invention or is nothing at all'.[6]

CONSTRUCTION OF MATHEMATICAL TIME

In a chapter of *Science and the Modern World*, Whitehead notes that two of Kant's assertions in the *Critique of Pure Reason* are contradictory or make time impossible. The first assertion concerns the mathematical continuity of time: 'I can only think in this portion of time the successive progress from one instant to the next, so that at the end, through all these parts of time and their addition, a definite quantity of time is reproduced'. The second affirms that time and space are continuous quantities, because there is no part that is not itself still space or time, bounded by points and instants. Whitehead sees this as an insurmountable contradiction, the first asserting that one must pass through the instant in order to conceive of time, and the second expressly rejecting the instant as a constituent of

5   A. N. Whitehead, *Process and Reality* (New York: Macmillan, 1929), 92. When the author speaks of 'these lectures', he means the Gifford Lectures, given in Edinburgh in 1927, which are the source of *Process and Reality*.
6   H. Bergson, *Creative Evolution* (London: Macmillan, 1907/1922), 361.

time. Thus, judges Whitehead, time is impossible if one adheres to both assertions. He accepts the second, but rejects the first.

> Realisation is the becoming of time in the field of extension. Extension is the complex of events, *qua* their potentialities. In realisation the potentiality becomes actuality. But the potential pattern requires a duration; and the duration must be exhibited as an epochal whole, by the realisation of the pattern. [...] Temporalisation is realisation. Temporalisation is not another continuous process. It is an atomic succession. Thus time is atomic (i.e., epochal), though what is temporalised is divisible.[7]

However an objection may come to mind: how can Whitehead say that concreteness does not take place in time, when the actual entity that is the product of it is born with a thickness of time, a 'duration'? It should be remembered that when he introduced this notion of duration, Whitehead had in mind the example of sound vibrations and the development of wave mechanics. A musical note cannot be characterised as a vibration of a given frequency until at least one complete period of vibration has occurred. The properties of an elementary particle cannot be defined as long as at least one period of vibration of its associated wave has occurred. In *Science and the Modern World*, Whitehead writes: 'Time is sheer succession of epochal durations. But the entities which succeed each other in this account are durations. The duration is that which is required for the realisation of a pattern in the given event. Thus the divisibility and extensiveness is within the given duration.

---

7   A. N. Whitehead, *Science and the Modern World* (New York: Macmillan, 1925), 129.

The epochal duration is not realised via its successive divisible parts, but is given with its parts'.[8] A little further on, he develops:

> It will be remembered that the continuity of the complex of events arises from the relationships of extensiveness; whereas the temporality arises from the realisation in a subject-event of a pattern which requires for its display that the whole of a duration be spatialised (*i.e.*, arrested), as given by its aspects in the event. Thus realisation proceeds *via* a succession of epochal durations; and the continuous transition, *i.e.*, the organic deformation, is within the duration which is already given.[9]

These last sentences probably deserve an additional explanation to be well understood. A concrescence occurs: it is the crystallisation of an actual entity, a piece of concrete reality, with its spatial extension and its duration. The succession of concrescences takes for us the appearance of a continuous flow that we call time. It is, as Whitehead insisted, a pure abstraction that the human mind makes from the concrete world. But this abstraction possesses mathematical virtues that the concrete reality does not possess: that of divisibility *ad infinitum*, or, if one prefers, of perfect continuity. One can, from there (and from a similar treatment with regard to space), construct the extensive continuum that physicists prefer to continue to call space-time. The continuous transitions of events in this abstract continuum are called 'motion' or 'evolution'; these are the 'organic deformations' Whitehead refers to in the previous quotation.

---

8   Ibid., 128.
9   Ibid., 137.

However, let us beware of deducing from this a kind of dualism of temporality, for which the time of the world would be one thing, and the time of our mind another, both of which being necessary to understand the world. Whitehead's philosophy, as we have seen with regard to the quantum theory and the interpretation of the collapse of the wave function, aims precisely at avoiding this rampant dualism, even expressly claimed by supporters of the Copenhagen School (such as Werner Heisenberg or Eugène Wigner). Abstraction is here only a mathematical tool, used to produce the concept of continuous time. Because for the Whitehead of maturity, the only concrete entities are the actual entities, born within the *process* that is the real engine of the world.

REAL TIME AND POTENTIALITY

Let us now try to clarify the relationship of time to potentiality. I believe that the majority of us will agree that the future cannot be inscribed in concrete reality, qualified itself as such. As long as it has not been realised, it remains in the limbo of potentiality. Only the proponents of the absolute determinism of events can contest this proposition,[10] but their position is desperate, in the true meaning of the word; Whitehead was certainly not the last to fight it. He even allows himself to ironise about it when he writes: 'This account of nature and of physical science has, in my opinion, every vice of a hasty systematisation based on a false simplicity; it does not fit the facts. Its fundamental vice is that it allows of no physical relation between nature at one instant and

---

10 As is well known, this point of view was embraced by Einstein himself, when on the death of his friend Michele Besso in 1955 he wrote to his family 'For people like us who believe in physics, the separation between past, present and future has only the importance of an admittedly tenacious illusion'.

nature at another instant. Causation might be such a relation, but causation has emerged from its treatment by Hume like the parrot after its contest with the monkey'.[11]

Of course, one should not exaggerate the significance of such a quotation. Whitehead is well aware that without causality there would be no permanence. But the mechanism he thinks of is radically different from that envisaged by classical science. As David Griffin (philosopher and director of the *Center for Process Studies* in Claremont, California) said: for Whitehead, 'time, or temporality is an ultimate feature of reality. It is not itself an actual or concrete entity; it is a *relation*—a relation of conformity to and inclusion of the past'.[12] We have explored this dimension of Whitehead's thought (and its proximity to Bergson's) in Chapter 6.

The emphasis on the potentiality of the future also has the advantage of diminishing the paradoxical aspects of quantum theory, as we saw in Chapter 7. In particular, it provides an elegant way of freeing oneself from the use of awareness or knowledge as a determining element in measurement processes. For Whitehead, decidedly, it is not the observer who kills or saves Mr. Schrödinger's cat![13] In *Process and Reality*, Whitehead

---

11 Whitehead alludes here to a very popular tale in America, where the owner of a monkey and a parrot, returning home after an absence, finds the monkey adorned with the coloured feathers of the parrot but can no longer find the parrot (A. N. Whitehead, *The Interpretation of Science, Selected Essays* (Indianapolis: Bobbs-Merrill Co., 1961), 57).

12 D. Griffin, "Time and the Fallacy of Misplaced Concreteness," in *Physics and the Ultimate Significance of Time*, ed. D. R. Griffin (Albany: State University of New York, 1986), 6.

13 In a thought experiment discussed at length by physicists and philosophers of quantum theory, a cat in a box is subjected to a non-deterministic quantum mechanism likely to kill it. For some theorists of the Copenhagen School, it is the observer, by opening the box and observing the cat, who determines its state, dead or alive.

writes, amongst other things, about the place of potentiality in the description of nature: 'It cannot be too clearly understood that some chief notions of European thought were framed under the influence of a misapprehension, only partially corrected by the scientific progress of the last century. This mistake consists in the confusion of mere potentiality with actuality. Continuity concerns what is potential; whereas actuality is incurably atomic'.[14] He adds: 'The alternative doctrine, which is the Newtonian cosmology, emphasised the "receptacle" theory of space-time, and minimised the factor of potentiality. Thus bits of space and time were conceived as being as actual as anything else, and as being "occupied" by other actualities which were the bits of matter. This is the Newtonian "absolute" theory of space-time, which philosophers have never accepted, though at times some have acquiesced'.[15]

REITERATION AND NOVELTY

Let's remember that for Whitehead, the reality of time, with its irreversibility, is based on the fact that the current world is only composed of momentary events that include, in part but in reality, previous events, which in turn include previous events, and so on.[16] It is this network of successive entities that articulate and partially merge into each other that is the source of temporal irreversibility. Whitehead explicitly states it: 'This passage of the cause into the effect is the cumulative character of time. The irreversibility of time depends on this character'.[17] If these successive handovers did not generate time, there would

---

14 Whitehead, *Process and Reality*, 61.
15 Ibid., 70.
16 Griffin, "Time and the Fallacy," 10.
17 Whitehead, *Process and Reality*, 237.

be no time in nature, because, as Milič Capek (another late Boston philosophy professor) pointed out, we must remember Hume's prophetic intuition, found in his *A Treatise on Human Nature* (book 1, part 3, section 2), when he points out that a total simultaneity between cause and effect would lead to 'the destruction of succession and the utter annihilation of time'.[18]

Naturally, when the chain of concrescences we are talking about produces natural entities that look very similar, or even such that some of their characters are identically preserved, it produces what we perceive and call 'permanent objects'. Why not qualify these permanent realities as *substances*, as Aristotle and classical physics invite us to do? Whitehead does not want this; he does not want us to lose sight of the fact that the permanent object is actually a succession of re-created entities, in which at any moment there is a chance, a potentiality of modification, however small. In *Science and the Modern World*, he asks: 'Has our organic theory of endurance been tainted by the materialistic theory in so far as it assumes without question that endurance must mean undifferentiated sameness throughout the life-history concerned? Perhaps you noticed that (in a previous chapter) I used the word "reiteration" as a synonym of "endurance". It obviously is not quite synonymous in its meaning; and now I want to suggest that reiteration where it differs from endurance is more nearly what the organic theory requires'.[19] A little further on, he clarifies his thinking. Permanence is indeed the product of the reiteration, in a series of successive concrescences, of a set of characters. But this miracle has more solidity than pure succession would

---

18 M. Capek, "The Unreality and Indeterminacy of the Future in the Light of Contemporary Physics," 303.
19 Whitehead, *Science and the Modern World*, 134.

lead one to believe, because 'It will be noted that endurance is not primarily the property of enduring beyond itself, but of enduring within itself. I mean that endurance is the property of finding its pattern reproduced in the temporal parts of the total event'.[20] Whitehead again alludes here to the fact that, at each concrescence, reiteration is not the re-creation of the new entity from nothing, although similar to the previous one. By means of the prehensions it weaves with the entities of the past, the re-created entity makes the structure of the past entity its own; this is what makes Whitehead, with Bergson, say that the present is continually enriched with the past.

On the other hand, let us remember that the net of prehensions that extends from past actual entities onto the concrescence in the making does not totally dictate what it will be, namely, a simple replication of the past. Reiteration does not exclude change. The simplest modifications concern movement: the form is preserved, but the successive concrescences occupy imperceptibly different locations, obeying the laws of conservation of the quantity of movement. Modifications of a higher level can also occur, and do indeed occur, bringing about what we call evolution. To explain evolution, Whitehead has recourse to the possibility, for the concreteness in the making, to select the past entities that it is going to prehend, to choose a number of them, many undoubtedly in the net of possible prehensions that converge towards it, but rejecting some of them, and perhaps also to draw from the reservoir of 'eternal entities' to lace with some of them prehensions that generate new and unexpected characters. This possibility of choice naturally implies what Whitehead calls a 'mental pole' of the actual entities, insofar as they express a preference, a striving towards an ideal, an aim (but note that this does not necessarily

---

20 Ibid., 153.

imply consciousness). As for the eternal entities, I see them as the mantle of the deity, its primordial nature, to speak as Whitehead; but we will come back to this in the last chapter.

It is when talking about perennial objects that one has the possibility of exercising a distinction between space and time. For perennial objects deploy a spatial structure that remains similar from concrescence to concrescence. This allows an observer to distinguish it, to follow its evolution, in short, to separate space and time for it. And for us, humans, to associate consequently to any perennial object (whether it is, e.g., another man or a clock), a time which in all rigour should be its proper time, the only one which exists according to the theory of Relativity. Whitehead describes and extends this situation in *Science and the Modern World*. He sees actual entities not as independent and separate entities, but rather as Leibniz-like monads, summarising also in them the globality of the present world and paying attention to their overall coherence.

He imagines the actual entities watching over each other with benevolence, ready to engender together the next actual entity. Thus, talking about individual entities is in reality an exaggeration or a borderline case. The current entities form societies, networks, or as Whitehead says, *nexuses*, a Latin word that evokes at the same time the temporal enchainment and the interlacing. Whitehead answers the objection that microscopic concrescences, those that concern what he calls event points, generally concern regions of space-time too small for an 'observer' to distinguish them and grasp their character of permanence. This is why, in *Science and the Modern World*, he insists on the global, societal character of permanent objects and structures: 'On the materialistic theory, there is material—such as matter or electricity—which endures. On the organic theory, the only endurances are structures of activity, and the structures are evolved'.[21] Patrick Hurley, another Whiteheadian professor of philosophy in the United States, explains: 'No one, of course, can *see* an actual occasion in the external world. What we see are nexus of occasions. A nexus in which a common element of form is causally transmitted from earlier to later members is termed a "society". Times makes its appearance as a feature of societies; namely, the feature by which societies endure'.[22]

DO THE LAWS OF PHYSICS EVOLVE?

In the last quotation we made of Whitehead, he lets us guess that he thinks that 'the structures of activity', that is, the laws of physics, must evolve with the evolution of the universe. This is an essential point of the philosophy of the organism, on which he will insist a lot in his later works, taking the risk of contradicting the physicists who, as a whole, marvel rather at the stability of the laws of the Universe. Already in *Science and the Modern World*, he states this new principle: according to the theory of the organism, 'the evolution of laws of nature is concurrent with the evolution of enduring pattern. For the general state of the universe, as it now is, partly determines the very essences of the entities whose modes of functioning these laws express. The general principle is that in a new environment there is an evolution of the old entities into new forms'.[23] What he affirms there seems indeed natural, within the framework of his organic theory of the universe. The structural laws of nature depend on the complex interactions that make up the

---

21 Ibid., 110.
22 P. Hurley, "Time in the Earlier and the Later Whitehead," 105.
23 Whitehead, *Science and the Modern World*, 109.

net of prehensions between actual entities, which are constantly renewing and changing, albeit little by little. Logically, the laws and what are today called the fundamental constants of nature (speed of light, electron charge, gravitational constant, Planck constant in quantum mechanics, fine structure constant in nuclear spectroscopy) could—should—have changed during the evolutionary phases of the universe. Otherwise, are we not reduced to thinking that these constants were fixed once and for all by the decree of a mind outside the universe? This is a real challenge for materialist or unirealist physicists. The field of research on the cosmic evolution of the laws of nature and fundamental constants has developed greatly in recent years. Many of these constants have been (indirectly) measured with increasing precision; as we saw in Chapter 8, these measurements have not detected any variation for billions of years. Perhaps this is why the majority of today's physicists, but not all philosophers of science, stay away from Whiteheadian thinking about time and space.

TIME AND CREATION

For Whitehead, physical laws 'exist as average, regulative conditions because the majority of actualities are swaying each other to modes of interconnection exemplifying those laws. New modes of self-expression may be gaining ground. We cannot tell. But to judge by all analogy, after a sufficient span of existence our present laws will fade into unimportance. New interests will dominate'.[24]

As a first intention, therefore, Whitehead admits that he would be tempted to adhere to a pessimistic vision of nature, probably a nature driven by chance framed by momentary

---

24 Whitehead, *Modes of Thought*, 212.

deterministic laws that change according to the general organisation of the world. But he immediately recognises that a
deeper reflection leads him to radically modify this vision. In
the depths of this general rule of chance and necessity, the social
organisation of things and beings reveals a codicil of the law, a
tendency, an 'appetite' he says, referring to Leibniz, a creative
thrust. And he insists on this hidden aim, on a final explanation
of the evolution of the universe, on the legitimisation of the
appearance of novelty, thanks to the selection of prehensions
between actual entities and the recourse to eternal entities. For,
he continues, 'it is untrue to state that general observation of
mankind, in which sense-perception is only one factor, dis-
closes no aim. The exact contrary is the case. All explanations
of the sociological functionings of mankind include "aim" as an
essential factor in explanation.'[25]

Before addressing the reasons which thus pushed Whitehead
to return, after a long period of open agnosticism, to reintro-
duce deity into the description of the evolution of the world
brought about by the process, let us now see how Whiteheadian
temporality poses the problem of the relations between determinism and freedom.

---

25 Ibid., 213.

CHAPTER 10

Free Will Saved from Determinism?

Whitehead's reflection on the subject of time highlighted the complexity of this common notion. For him, this complexity has not been sufficiently translated into physical theory, whether it is Newtonian time or even Einstein's relativistic time. Placed at the heart of the organic vision of the world, the Whiteheadian notion of *process* encompasses both a chopped image of the world and its future, through the successive concrescences that are indivisible acts, and a smooth time woven by permanent things, perceived by the human mind. However, this complex texture has a great merit: by refusing a continuous motor at the heart of evolution, it breaks absolute determinism or at least relativises it. We will now see that it sheds new light on the problem of the relations between body and mind, and more precisely on the very embarrassing question in philosophy of free will in the face of the determinism of the bio-electrical activities of neuronal circuits in the brain.

To see this, we now need to take a trip into the field of neuroscience. In this young discipline (the word 'neuron' is hardly more than a hundred years old), we find many researchers convinced that the central nervous system, like any other system subject to 'inputs' from the outside world and providing one or more 'outputs', whether it is itself mechanical or a living organism, automatically transforms these inputs into outputs, even though it may draw from its memory certain characters of the output produced. There are others, and I confess that I am one of them, who claim to be 'emergentists', that is to say, they do not exclude the possibility, at least in certain circumstances, of unpredictable deviations from this beautiful regularity (but depressing automaticity) predicted by the former.

Let us take the particularly telling case of Roger Wolcott Sperry, a twentieth-century American neuropsychologist. As a young student, he received his first training in neuropsychology in a laboratory immersed in the influence of the then-dominant deterministic 'behaviourist' philosophy. However, it was his own observations—and this is particularly remarkable—that gradually led him to propose a typically emergentist vision of the brain function. His work earned him the Nobel Prize in 1981.[1]

At the age of 21, Sperry was awarded a scholarship to study psychology at the University of Oberlin, more precisely Oberlin College. His main teacher was Raymond H. Stetson, and after receiving his Bachelor of Arts degree, he remained in the latter's laboratory for another two years as an assistant. Professor Stetson studied speech from the point of view of the muscle movements necessary to produce it. As I have already mentioned, the atmosphere of psychological research at that time was strongly influenced by what has been called 'behaviourism', especially following the publication in 1924 by John Watson of the book of the same name. Behaviourism professed that

---

1   In presenting the observations of this great scientist who is a little forgotten today, I will mainly draw inspiration from an article (R. Lestienne, "Emergence and the Mind-Body Problem in Roger Sperry's Studies," *Kronoscope* 13 (2012): 112–26) that I have already published on this subject, as well as from my book *Dialogues sur l'Emergence*. R. Lestienne, *Dialogues sur l'Emergence* (Paris, France: Le Pommier, 2012); *Dialogues about Emergence, Kronoscope* 16 (2016): 15–135.

the brain was too complex an organ to understand its internal functioning. It was therefore necessary to look at it as a 'black box' and to be content with studying the correlations between inputs and outputs; for example, between the instruction 'produce an "a"' and the consecutive muscular movements of the face. The research in Stetson's laboratory was precisely on this and similar topics. He had equipped his laboratory with instruments for recording the finest muscle movements and observing their coordination, and taught Sperry how to use them.

CHANGING THE DESTINATION OF PERIPHERAL NERVES TO OBSERVE CHANGES IN MOTOR BEHAVIOUR

In 1937, Sperry left Oberlin for Chicago, where he worked for four years under Professor Paul Weiss and obtained his doctorate in 1941. The atmosphere in Paul Weiss' laboratory was quite different: Paul Weiss was a pure biologist, specialised in embryology and animal development. Under his guidance, Sperry quickly became a very gifted experimenter who was able to perform a delicate operation on various animals: nerve transposition. In particular, he performed in young rats under anaesthesia the reconnection of the sensory nerve of the right hind leg to the ascending sensory nerve of the left hind leg, after sectioning the latter. Thus, in these animals, once the sutures had healed, the sensory nerve of the right paw now ascended towards the right hemisphere of the brain, whereas in normal animals it ascends towards the left hemisphere (the crossing of the sensory and motor nerves at the entrance to the brain is a general rule in mammals, including humans).

However, this non-crossing of nerves had surprising consequences on the later behaviour of the adult animal. When a slight shock was applied to the animal's right leg, it always lifted the other leg (the left), despite the fact that it could see what was being done to it (see Figure 10.1). For Sperry, this was a clear indication that the association of sensations with a limb was controlled once and for all by the central nervous system (the brain), and not in a way that could be modified by experience. He later explained that 'the main point [...] is the contention that the animal responses on protectively holding up the wrong foot and in yipping and licking the wrong foot are caused directly in brain function by the subjective pain property itself, rather than by the physiology of nerve impulses'.[2]

Sperry's doctoral thesis, defended in 1941, was entitled "Functional Results of Crossing Nerves and Transposing Muscles in the Fore and Hind Limbs of the Rat". It should be noted that the protocol followed was typically a behaviourist one: stimulation of the rat's leg and observation of the subsequent motor behaviour, notwithstanding of course the episode of the procedure performed: this remained limited to the peripheral nervous system. Despite Paul Weiss's own open-mindedness, at that time Sperry remained imbued with the behaviourist spirit. Even as late as 1952, when he published a detailed account of his experiments in the *American Scientist* magazine, although he already marked a certain distance from the theses of behaviourism, he still let himself write that *the entire output of our thinking machine consists of nothing but patterns of motor coordination*, and that *the entire activity of the brain, so far as science can determine, yields nothing but motor adjustments*, and again that *any separation of mental and motor processes in the brain would seem arbitrary and indefinite*. These are, of course, typically behaviourist sentences. The reason for

---

2   R. Sperry, "Mental Phenomena as Causal Determinants in Brain Function," *Process Studies* 5 (1976): 247–56.

this conciliatory attitude towards behaviourism was probably his conviction that the same percepts, thoughts, and feelings, could correspond to very different patterns of nerve activity in

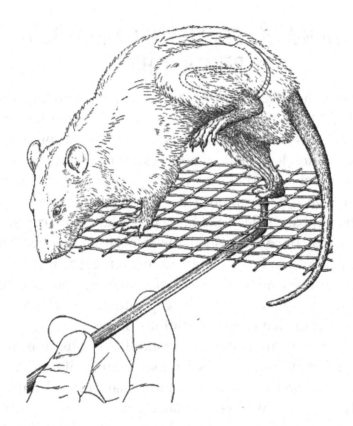

Figure 10.1. *Giving a small shock to the hind limb of a rat that has had its sensory nerves in the hind legs reversed, the animal responds by invariably lifting the other leg. Sperry concludes that the association of pain with a limb, once acquired and stabilised, is acquired forever by the central nervous system, and is no longer modulated by experience. Original drawing by Sperry.*[3]

---

3   R. Sperry, "The Growth of Nerve Circuits," *Scientific American* 201 (1959): 68–75.

the various associative areas of the brain, but that these stimuli must necessarily converge towards the same precisely defined postures and movements. This was clearly one of the favourite arguments of behaviourism.

PAIN AS A KEY TO TESTING MIND–BODY RELATIONSHIPS

Sperry, however, no longer fully agreed with this theory. He felt that behaviourism's refusal to explore the territory of consciousness was exaggerated. Nerve-crossing experiments in rats, soon supplemented by other experiments in various animals, undoubtedly showed the innate character and strength of associations between one side of the central nervous system and the opposite side of the body or visual field. They seemed to show that, once established, the association between a nerve and a part of the body remained forever, even in the case of surgical transposition of peripheral nerves, and despite the fact that the animal seeing the experimenter's gestures could observe the inadequacy of its motor response!

A first step to better understand the situation and go beyond behaviourism was to use what philosophers call *qualia*, that is, subjective sensations that cannot be analysed simply in terms of physical causes. A common example of qualia is the colour red. This colour as felt is not only a property of the object we are looking at, but a quality of sensation that depends both on the physical properties of reflection of light on this object and on the sensory processing carried out in our brain.

But the colour red is not Sperry's favourite example. Perhaps because of the type of experiments he was performing in animals or for other reasons, he chooses to reflect on the sensation of pain. He was convinced that pain was not reducible to the excitation of specialised nerves or specific sites in the brain, but rather thought that this feeling was related to a particular structuring of brain excitations in time and space. There is nothing strange about this view. Think of the pain sometimes keenly felt by patients who have had an arm or leg amputated: they are actually suffering from their missing limb, the hand on that arm or the toe on that leg that they have lost. This observation and others like it have persuaded Sperry that pain cannot be reduced to the biophysical or physiological causes one can think of. Nerve excitations are multivalent: nerve impulses can generate sensations of pleasure or pain, depending on the networks of excited neurons and their spatiotemporal structure. However, these structural features are normally determined by the excitation of peripheral pain receptors, which do not exist in the case of amputated limbs. In an amputee patient, they are therefore created from scratch by the brain. Sperry felt that this argument was decisive.

PASSING FROM ANIMAL PERIPHERAL NERVES TO THE HUMAN BRAIN

Peripheral nerve transposition experiments could not completely satisfy Roger Sperry. As doctors are well aware, leg reflex responses, for example, do not necessarily mobilise the brain, they are often initiated at the spinal cord level. After integrating the *California Institute of Technology* (CALTECH) in Pasadena in 1954, Sperry began a series of experiments to sever the corpus callosum (the rich bundle of nerve fibres through which the left and right hemispheres of the brain exchange information) in cats and monkeys.

In 1962, he had the opportunity to discuss his research with Dr. Joseph Bogen who was also working at CALTECH. The latter had been visited some time earlier by a 43-year-old World War II veteran who had suffered a wartime head injury from shrapnel and since that time had been suffering from severe epileptic seizures that could no longer be controlled by medication. Dr. Bogen had introduced this patient to neurosurgeon Philip Vogel, and in agreement with the patient, they decided to surgically separate his right and left brain hemispheres completely, in the hope of preventing an epileptic seizure born in one hemisphere from spreading to the other.[4] This was not the first such procedure, but it was relatively rare. In this patient's case, the intervention was successful: he had maintained his high IQ and the seizures had virtually disappeared. Sperry immediately understood the value of this unique opportunity. He requested permission to interview the patient and, with his consent, to submit him to some psychophysical tests.

He organised these tests in his own laboratory, with the help of a young neuroscientist, Dr. Michael Gazzaniga. Later, he was able to extend this type of research to about twenty patients who were also commissurotomised.[5] A schematic diagram of the experimental conditions, taken from an original drawing by Sperry, is shown in Figure 10.2.

The patient is seated at a table equipped with an opaque screen (oblique on the image) that prevents him from seeing his hands and the objects he can explore by touch, as well as a

---

4   This operation requires sectioning not only the corpus callosum but also two other small bundles of communication between the hemispheres, the anterior commissure and the posterior commissure. However, the optical chiasma is left untouched in order to preserve the entire binocular vision.

5   Commissurotomised: whose commissures linking the two cerebral hemispheres have been severed.

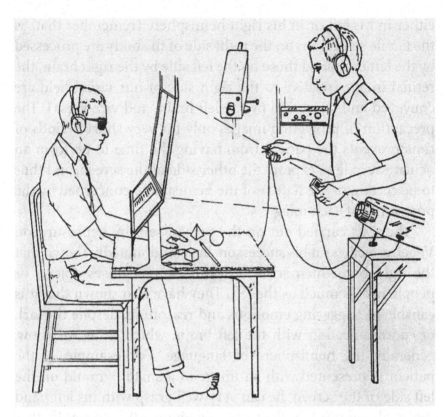

Figure 10.2. *Experimental arrangement to identify and study symptoms following surgical separation of the two hemispheres of the brain, according to Sperry.*[6]

translucent vertical screen on which images can be projected, in front of his eyes. The images are usually projected for brief moments (less than 100 ms), sometimes on the right side of the screen, sometimes on the left side, while the patient is asked to stare straight ahead. Thus, the images are processed separately,

---

6   R. Sperry, "Perception in the Absence of the Neocortical Commissures," *Perception and Its Disorders* 48 (1970): 123–38.

either in his left or in his right hemisphere (remember that, as the tactile sensations on the right side of the body are processed by the left brain and those on the left side by the right brain, the retinal images relative to the right side of our visual field are conveyed and processed by the left brain, and vice versa). The precaution of projecting images only for very short periods of time prevents the patient from having the time to perform an ocular saccade to explore the other side of the screen, and thus to be certain of the nature of the hemisphere concerned by the processing of each image.

The tests carried out on the person sent by neurosurgeon Vogel as well as on his successors have undoubtedly shown that the right brain memorises, categorises, recognises objects or people just as much as the left. They have also shown that it is capable of triggering emotions and reasoning, despite the lack of communication with the left brain, which is, as we know, generally the hemisphere of language. For example, if the patient is presented with an image of a small pyramid on the left side of the screen, he can very well grasp with his left hand the small pyramid on the table (see Figure 10.2) to signify that he has understood what object was presented on the screen. The image does not need to be a faithful reproduction of the object he will choose: the right brain is capable of reasoning by association. If the image of a nail is presented on the left side of the screen, he can very well choose with the left hand on the table the small hammer that would have been placed there amongst other objects. However, in all these cases of image presentation in his left visual field, due to a lack of information exchange between his right and left hemispheres, he is unable to *say* what he saw; indeed, in the vast majority of patients, the language centres are located in the left hemisphere.

Sperry and his team were also struck by the fact that in commissurotomised patients, consciousness, or more precisely the unity of the reconstructed perception of the world and of oneself, was nevertheless perfectly preserved. There was no sense of the oddity of their situation. Any discrepancy between what was shown to them on the left side of their visual field and what was shown on the right side was immediately reinterpreted in a coherent framework. For example, if, out of sight, they were pinched or pricked on their left hand, they would wince, but they could not tell where their pain was coming from. If they are questioned, they interpret their pain in such a way as to protect the coherence of their vision of the world, even if it means making up stories. Yes, to fabricate, that is, to invent reasons for their behaviour instead of reasons to which their left brain does not have access. They will say, for example, that they wince because they have heartburn. Or, if they are shown a naughty image in their left visual field, they will blush. If we ask them why they blush, they will say, for example, that the laboratory is too hot, or that they are prone to hot flashes, and so on.

The many such observations led Sperry to look at consciousness as something unique, beyond mere neuronal excitations. Sperry himself notes: 'My long trusted materialist logic was first shaken in the spring 1964. [… From September 1965,] I openly changed my alignment from behaviourist materialism to antimechanistic and non-reductive mentalism (as the term "mentalism" is used in psychology in contrast to behaviourism; not, of course, in the extreme philosophic sense that would deny material reality).'[7]

---

7   R. Sperry, "Mind–Brain Interaction: Mentalism, Yes; Dualism, No," *Neuroscience* 5 (1980): 195–206.

THE UNITY OF CONSCIOUSNESS

Thus, for Roger Sperry, the main reason for considering consciousness as an emergent global entity with its particular special properties is the unity of consciousness, not only in normal subjects but even in commissurotomised subjects. The first thing that surprised Sperry and his team in examining these subjects was the absence of visual symptoms in ordinary life, especially in the median zone, where the right and left visual fields—processed by two independent brain hemispheres—connect. One can very easily persuade oneself of this. When you close one eye, the part of your visual field facing the closed eye is processed by the other hemisphere than the one facing the open eye. Of course, in normal subjects, the treatments by both hemispheres are corrected and merged thanks to the communication channel constituted by the corpus callosum. As Roger Sperry confirms, commissurotomised patients do not complain about visual problems, 'nor do the patients' comments indicate that they are bothered by or even notice any changes or peculiarities in their visual experience. There is no reference, for example, to any halving or doubling in their vision, nor to any imperfection or irregularities at the vertical midline [separating the two visual hemifields]'.[8]

***

Sperry was thus led to admit that consciousness exerts a causal, reorganising power on the underlying neural activities. For him, the perfect unity of consciousness seems unthinkable without this organising power, which perhaps would be more accurate to qualify as a bifurcation power: consciousness, faced

---

8   Sperry, "Perception in the Absence of Neocortical Commissures."

with inconsistencies, seems capable of re-orienting, of provoking bifurcations in the functioning of this or that neural network, whether it is to restore the failing coherence of sensations and the unity of the self, or to obey an impromptu voluntary mental decision.

One can wonder about the nature of this intervention of an 'impromptu' voluntary decision which would favour a bifurcation in the course of information processing in the neural circuits of the brain: is this not the return of a form of Cartesian dualism? The objection was formulated to him, notably by two of his own friends, Karl Popper and John Eccles: both, as we know, brilliant minds,[9] and both in favour of the specificity of thought in relation to matter. In 1977, these two authors published *The Self and Its Brain*, in which they quoted extensively from Sperry's work, which they interpreted in favour of a dualistic vision of the mind–body problem. For them, in fact, the power of consciousness is so strong that in its very nature it must necessarily be separated from physiological processes. Consciousness is what selects, from the neural networks and the organised structures of their arousal, those networks and structures which together compose the unitary and conscious vision of the subject. Moreover, because according to them, consciousness is totally dependent on language, they believe that the notion of 'being conscious' must be circumscribed and reserved for the left brain only (in right-handed people).

Roger Sperry feels obliged to answer. He does so in a long article entitled "Mind–Brain Interaction: Mentalism, Yes; Dualism, No".[10] Of course, he refused this hierarchy between

---

9   Karl Popper was one of the most brilliant philosophers of science of the twentieth century. John Eccles was awarded the 1963 Nobel Prize for discovering inhibitory neurons in the brain.

10 Sperry, "Mind–Brain Interaction."

the two hemispheres, because his observations on commissu-rotomised patients had shown him, on the contrary, the equal mental power of the two hemispheres. In a previous article, he had already stated his conviction that his observations did not require him to renounce either the monism of thought or his biological nature: mental events are not correlates of nerv-ous excitement, the latter are the cause of the former. There is therefore a kind of interaction between thought and neural processes, but this interaction implies neither dualism nor par-allelism. Mental forces act as an emergent cause in neural pro-cesses.[11] In his 1980 response to Popper and Eccles, he develops this argument in favour of monism and, against these authors, maintains the determinism of neural events.

At this point, let us summarise Sperry's position by review-ing the four main arguments on which he bases his emergent view of consciousness. First, there is the specificity of *qualia*, in particular that of pain. The pain felt by an animal whose sensory nerves in the lower limbs have been crossed is not caused by the sensory receptors of the leg it is lifting, which he nevertheless unambiguously identifies as concerning *that* leg. The pain an amputee patient suffers from is not caused by his lost hand or leg either, but by a ghost that exists only in his brain. 'Pain as a subjective experience was explained [by me] as a holistic property of a particular spatiotemporal pattern of cerebral excitation that as a dynamic functional entity directly determines the further course of brain activity.'[12]

Then there is the integration of the perceptions experienced as a whole, rather than a discontinuous sequence of sensations.

11 Sperry, "Mental Phenomena as Causal Determinants."
12 R. Sperry, "An Objective Approach to Subjective Experience: Further Explanation of a Hypothesis," *Psychological Review* 77 (1970): 585–90.

One of Whitehead's favourite examples is perceptual images of the objects around us. Sperry joins him on this point:

> As you look about the room it is the images you see that the brain process is responding to directly, rather than to the nerve impulse components and other subelements of which these same conscious images are built. The latter point is critical, namely, that the brain responds to the overall encompassing effect or functional gestalt[13] of an excitation pattern as an entity, rather than to the individual impulse elements of which the excitation process is composed.[14]

Another example is the melodic unity of a grouped series of musical notes (here we find William James' 'thick present'):

> Consider the perception of a melody whistled or tapped on a piano. [...] The musical notes of subjective experience are part of the brain activity itself and are located within the brain. [...] The overall pattern of this excitation hierarchy is presumed to possess, in proper context, the experienced subjective properties with their causal effects that enable such an excitatory pattern to be treated in cerebral function as a unit. The subjective experience is conceived to be one among a large variety of holistic properties with causal effects in the brain process existing and operating at different levels of organization.[15]

---

13 Sperry refers here to the German school of gestalt psychology, which emphasises the importance of functional sets of perception (form, objects) rather than their constituent elements (segments, details).
14 Sperry, "Perception in the Absence of Neocortical Commissures."
15 Sperry, "An Objective Approach to Subjective Experience."

Third, there is the banal experience of the perfect continuity of the visual experience between the right and left visual fields, which we have already thought about at length, despite the fact that these visual fields are processed in distinct cerebral hemispheres. This visual continuity could be explained by the exchange of information between the two hemispheres via the corpus callosum, but we have seen that this evenness is not destroyed by the complete section of the latter.

And finally, there is the coherent unity of consciousness, the idea we have already discussed about the corrections made by the brain to our perception of the outside world, or the coherence of conscious perception in commissurotomised subjects, despite the possible discordances of the images presented to them in their two visual fields.

We are thus in the presence of a monistic theory of the brain and the mind, on a par with Spinoza's monism, but without appealing to a double nature, a double attribute of the thinking thing.[16] A theory close to materialism, while denying that

---

16 'It is true that Spinoza's philosophy has a strong influence on today's scientists', says Sagredo (a shrewd researcher, who meets and dialogues with great scientists who are now deceased). 'You know, Spinoza was not satisfied with Descartes' dualism, which he had studied closely. His philosophical masterwork, *Ethics*, was published after his death, in 1677. There, he developed the reasons for his dissension with Descartes. In short, for him there is in the world only a single substance, reality, nature, which one can also call God. There cannot thus be two substances, "extension" and "thought," as people said during the Renaissance, or the "brain" and the "mind," as we would say today. Again, in contemporary language, the neutral monism of Spinoza suggests that the physiological processes in the brain and the contents of our consciousness are only two ways of speaking about the same thing, two modes that this single substance has to show itself to us under two distinct "attributes." Thinking substance and extended substance are one and the same substance, comprehended now under this attribute, now under that. There cannot be an interaction of the one on the other, no more, say, than

the mind is totally subject to the determinism of the underlying neural circuits. Sperry denies any reality to any living substance that is not made of biological cells. But his conception of determinism gives way to a part of emerging determinism, which appears only in a global system and exerts its action on its components. For him, 'of all the questions that can be asked about consciousness, none is as important, none offers deeper implications than whether or not consciousness plays a causal role. The diverse answers that can be given to this question lead to the whole panorama of philosophical, scientific and cultural systems [that we know]'.[17]

This is a courageous position, but not a comfortable one. Descartes' dualism, like that of Popper and Eccles, has the advantage of positing a difference in nature between the mind and the brain such that it allows us to conceive that the mind can exist outside matter. This one could thus survive death. This is clearly a point that most of the great religious traditions absolutely insist on. Roger Sperry's emergentism does not leave much hope of this kind, and moreover, it is a hypothesis that he himself has always vigorously rejected. For him, brain death necessarily leads to the disappearance of consciousness and all mental qualities.

---

between the two faces of a coin. Yet this image does not do justice to the total otherness that Spinoza recognises between thought and extension. This incapacity to interact is obviously incompatible with the emergentist viewpoint of Roger Sperry, for whom the possibility of a causal action of consciousness on the physiological processes of the brain is essential. It is, moreover, as difficult to reconcile with the reductionist or interactionist viewpoint of many contemporary neuroscience specialists, even among those who readily claim to adhere to Spinoza!' Excerpt from: R. Lestienne, "Dialogues about Emergence," *Kronoscope* 16, no. 1(2016): 15–135.

17 Sperry, "Mind–Brain Interaction."

RESTORING FREE WILL?

Do these elements for an emergent vision of consciousness, this causal power of the global brain on the underlying neural processes argue in favour of free will? Sperry believes it and claims it. 'In any decision to act, these conscious mental phenomena override and supersede the component physiological and biophysical events involved in the causal progression of brain activity'.[18] This does not mean that one can do anything, make any decision. 'Free acts are rare', Bergson once observed. Sperry echoes him: 'What we want from free will is not to be totally freed from causation but, rather, to have the kind of control that allows one to determine one's own actions according to one's own wishes, one's own judgment, perspective, cognitive aims, emotional desires, and other mental inclinations. This, of course, is exactly what is provided in our current interpretation [of brain function]'.[19]

This causal power alleged by Sperry and other proponents of the 'strong' emergence of consciousness seems paradoxical, however. If consciousness is indeed only the product of underlying nervous excitations that are themselves the necessary result of a deterministic process, how can it have a causal action on these same excitations and disrupt their course? Is this rejection of dualism, this monism of active thought and nervous activity, which Sperry claims so highly, coherent? Yes, argues the American neuropsychologist, because my conscious activity is an emergent, new, creative property that only appears in a brain that has crossed a threshold of organic complexity, an action of the whole on the parts that is unpredictable as long as

---

18 R. Sperry, "Changing Concepts of Consciousness and Free Will," *Perspectives in Biology and Medicine* 20 (1976): 9–19.
19 Ibid.

we look at the functioning of a subsystem, whose evolution is regulated by the laws of physics and biology. Is the reasoning defensible? Philosophers have grasped the question, especially those who have shown themselves, in recent decades, sensitive to emergentist arguments.

PHILOSOPHICAL CRITIQUE OF EMERGENT CONSCIOUSNESS

Jeagwon Kim is one of the contemporary philosophers who felt most concerned by the question, partly driven by the hope of a possible answer in the line of emergentism. This American–Korean philosopher, professor emeritus at Brown University in the United States, obtained his doctorate at Princeton in 1962, focusing on the notion of explanation in physical theory. Very quickly, he became interested in the theory of psycho-neuronal identity, and expressed his reservations about this theory:

> I conclude, therefore, that the adherents of the Identity Theory can find no support in the considerations of simplicity or unity in the structure of scientific theory. [...] But a radical Dualism [...] is not the only alternative to the Identity Theory. A sort of Dualistic Materialism results if one accepts the identity of the particulars involved in the two events but not the identity of the properties. But we will do well to remind ourselves that the economy [of thought gained by such a theory] would have to be attained in the face of the extreme implausibility besetting the factual identification of mental properties with physical ones.[20]

---

20 J. Kim, "III. On the Psycho-Physical Identity Theory," *American Philosophical Quarterly* 3, 1966: 227–35.

This discreet appeal to an emergentist theory of consciousness, however, receives a somewhat disillusioned echo forty-four years later by the same author. 'Lately, the emergence concept has, it seems, created a kind of bandwagon effect, engendering high enthusiasms and expectations', a statement immediately corrected by 'I should say up front that what I am going to say will by and large be deflationary if not negative, and I have more questions than I have answers'.[21] In a word, for the idea of strong emergence to be credible, the emergent property must indeed have a causal power of its own, what emergentists call a *top-down causality*. For if emergent property has no causal power, it has no power at all, and then it is very likely to appear as a simple epiphenomenon, an idea that could be avoided. Thus, if the principle of top-down causality were found to be incoherent, it would be a serious blow to emergentism. Now, as we shall see, Kim thinks that this principle is incoherent, because of the transitivity of the relations of sufficient causality.

To begin with, Kim proposes a clear definition of emergence, as the appearance of a radically new 'supervening'[22] state when a certain sequence of states of the subsystems N_1, ..., N_n is realised, but which is not reducible causally to them:

> Property M is emergent from properties N_1, ..., N_n only if (1) M supervenes on N_1, ..., N_n and (2) M is not functionally reducible with N_1, ..., N_n as its realizers. [...] I believe

---

21 J. Kim, "Being Realistic about Emergence," in *The Re-Emergence of Emergence. The Emergentist Hypothesis from Science to Religion*, ed. Philip Clayton and Paul Davis (Oxford: Oxford University Press, 2006), 191.

22 The 'supervenience' of an emergent property M refers to the set of conditions necessary for the M-state to be achieved. For example, M is a mental state that can occur, provided that the neural networks N_1, ..., N_n are realised and properly excited.

that the two clauses of Emergence capture the concept as it was introduced by the classical emergentists like Samuel Alexander, Lloyd Morgan, and C.D. Broad.[23]

If the emergent state M occurs when neural networks are activated in the states $N_1, ..., N_n$, but M has the immediate power to bring about a modification of these states, then it seems that it has the power to influence itself, a self-reflexive causality, which seems contradictory. 'How is it possible for the whole to causally affect its constituent parts on which its very existence and nature depend? If causation or determination is transitive, doesn't this ultimately imply a kind of self-causation or self-determination—an apparent absurdity?'[24] And he adds: 'One avenue still seems open: that the effects of emergent properties be understood as not acting at the same time as the causalities of the lower levels, in short, that their effects are staggered in time'. I have called this a 'diachronic top-down causality'.

This idea would thus require a causal, and therefore temporal, break in the evolution of the system under consideration. If we take again the example of thinking and the underlying neurophysiological processes, this would mean that thinking can only have its own effect on them in a diachronic way, with a temporal break. This may perhaps shock biologists, but not necessarily physicists: have the theorists of chaos and complexity not provided plausible explanations for the occurrence of bifurcations in the evolution of complex systems? And doesn't

---

23 Kim, "Being Realistic about Emergence," 197. Functionalism consists in characterising a property or explaining it in terms of the causal work it performs. For example, a gene is characterised by its causal function of transmitting phenotypic traits from parents to children.
24 J. Kim, "Making Sense of Emergence," in *Emergence, Contemporary Readings in Philosophy and Science*, ed. M. Bedeau and P. Humphreys (Cambridge: MIT Press, 2008), 146.

quantum mechanics teach a similar break in the collapse of the wave function?

But how can one claim that consciousness influences, although by simply provoking 'diachronic bifurcations', the course of the physiological processes of the brain and at the same time admit that these same processes are determined by the laws of physics and biology? This seems quite difficult to understand and to admit. Jaegwon Kim himself acknowledges it: the transitivity of sufficient causes continues to apply to this case, if we admit what he calls the 'causal closure', according to which causes are necessarily restricted to the physical domain.[25]

Let us stop for a moment, to better see where the problem lies. Let's take the example of an old-fashioned television set with its cathode ray tube. The flow of electrons which sweeps the screen by providing an image is precisely regulated by the laws of electromagnetism: the place of the impact and its intensity on the screen depends on the voltage at the terminals of the capacitors at the exit of the electron gun and the intensity of the current in the magnetic coils. But the coherence of each image and their succession is the result of the program planning. And the passage from one image to another, like the change in the content of our consciousness, does not violate any law of

---

25 Kim explains: Suppose that the mental property M is causally efficacious with respect to physical property P*, and in particular that a given instance of M causes a given instance of P*. Given the Physical Realisation Thesis, this instance of M is there because it is realised by a physical property, say P. Since P is a realisation base for M, it is sufficient for M, and it follows that P is sufficient, as a matter of law, for P*. Now, the question that must be faced is this: What reason is there for not taking P as the cause of P*, by-passing M and treating it as an epiphenomenon? (J. Kim, "The Non-reductivist's Troubles with Mental Causation," in *Emergence, Contemporary Readings in Philosophy and Science*, ed. M. Bedeau and P. Humphreys (Cambridge: MIT Press, 2008), 440).

electricity. To be sure, these changes are the result of an external intervention, the program planning in the television studio, and the modulation of the waves that fall on the television antenna. It can be said that in the case of the brain it is the overall state of the brain, in that it has brought about consciousness and its power of coordination, which plays a role similar to the program planning in the case of television programs. It will be objected that program planning does not belong to physical or biophysical causes, but for Sperry, it is the weakness of the call to an image. The will is indeed biophysical.

But a more direct solution seems to be offered by the Whiteheadian philosophy, that of accepting an interruption of temporal continuity in the process in question, in an emergentist and diachronic approach to the relationship between consciousness and the brain. After all, this would not be the first time that science has considered an interruption of continuity in temporal succession. We have seen the case of the collapse of wave function in standard quantum theory, which states, let us recall, that in any measurement process, the deterministic sequence of states is broken.

Figure 10.3 compares the situation as discussed by Kim with the situation reformulated within the Whiteheadian philosophy. The neural state N is the basis of the emergent state C of consciousness, which 'supervenes' or can occur when the neural base N is realised. Note that N is not strictly speaking the cause of C, insofar as or by hypothesis of emergence, C is not functionally reducible to N, but in the last analysis N is a sufficient condition for the occurrence of C. Emergentists claim that, by a descending causal effect, it is the cause of the modified neuronal state N^*, which is the basis of the new state of consciousness C^*. Kim reasons by observing that there is no reason not to consider N as the cause of N^*, avoiding the sufficient chain of causality between N and C and C and N^* (Figure 10.3A).

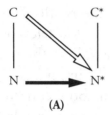 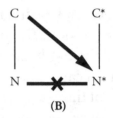

Figure 10.3. *The causal chain alluded to from the emergent state C of the global brain to the modified state N* of the subjacent neuronal circuits may be an illusion in an emergentist classical science (A), but be justified in the frame of a Whiteheadian philosophy of time (B).*

But for the proponents of Whitehead's philosophy, the causal link between N and N* can be broken, because of the discretisation of the concrescences between the states N and N*; on the other hand, the causal chain from C to N* can be functional, because of the mutual 'prehension' of N* and C.

This is the situation within Whiteheadian philosophy. The discrete nature of the concrescences of the N and N* states of the brain allows us to consider that the C state of consciousness is really the cause of the N* state of the brain. The emergent properties of consciousness thus seem to allow a direct causal effect on the underlying activity of neural networks in the brain. Whitehead would probably have said that the base of concrescence of N is narrower than the base of concrescence of N*, which necessarily encompasses C, unlike N.

I have outlined an extrapolation of Whitehead's philosophy regarding the old and painful problem of free will. Whitehead himself was not fortunate enough to know Sperry's work: his first publication in a widely read newspaper appeared in 1952, five years after Whitehead's death. Conversely, while Sperry was familiar with the work of the founders of modern emergentism (Conwy Lloyd Morgan, Samuel Alexander, and Charles Broad), he was apparently not familiar with Whitehead's books. In this chapter, I have tried to convey the immense scope of the philosophical denial of the reality of instants of time and the complex structure of nature's passage. The notion of free will is indeed central to metaphysics, as Roger Sperry had already reminded us: 'Of all the questions one can ask about conscious experience, there is none for which the answer has more profound and far-ranging implications than the question of whether or not consciousness is causal. The alternative answers lead to basically different paradigms for science, philosophy, and culture in general.'[26]

---

26 Sperry, "Mind–Brain Interaction," 205.

CHAPTER 11

Whitehead's God

As we know, during all the years of his studies at Cambridge, Whitehead, as the worthy son of an Anglican pastor, showed himself to be a convinced Christian; later, he moved away from religion, settling his long hesitation between the various Christian churches for agnosticism, or even the atheistic materialism shared by several of his friends, the first being Bertrand Russell. In 1897 or 1898, he declared his choice for agnosticism, became a militant of the cause, and sold most of the numerous books he had gathered on religious subjects. His wife Evelyn, not very interested in these subjects, was delighted with his decision. But, as his son Thomas North pointed out, throughout this long period of emerging agnosticism, his father's agnostic activism did not hide a kind of latent dissatisfaction, in spite, or because of the many criticisms he had made of the established churches.[1] In particular, from the beginning of this period, he vigorously contested the notion of an almighty God, creator of heaven and earth, which he felt was not compatible with the explanation of evil on earth. He also criticised the role and character of Saint Paul, who, in his opinion, had distorted the original Christian message. Lucien Price, a great Boston journalist, became after 1932 one of the most frequent

---

1   See especially V. Lowe, *Alfred North Whitehead, the Man and His Work, Volume I: 1861–1910* (Baltimore: The John Hopkins University Press, 1985), 188–90.

visitors to the Whitehead house and confided most of his con-
versations with the philosopher in his book *Dialogues of Alfred
North Whitehead*. He recounts several of Whitehead's digs at
the Apostle to the Gentiles: 'The trouble with the Bible has been
its interpreters, who have scaled and whittled down the sense
of infinitude to finite and limited concepts, and the first inter-
preter of the New Testament was the worst, Paul'; 'Jesus was
not very intellectual; what he had was a profound intuition.
Humanity in the Eastern Mediterranean between 500 B.C. and
200 A.D. began to write down their intimate thoughts and a
great age resulted. I am speaking of course of the exception-
ally gifted men who wrote down their thoughts. Paul comes as
quite a drop from Jesus, and although his followers included
many estimable persons, their idea of God, to my mind, is the
idea of the devil'.[2] Apparently he was thus reproaching Paul for
the invention of hell and the dogmatic skeleton of theology,
but is really Paul the author of these things? Whitehead also
denigrated in the Catholic Church the recent dogma of pon-
tifical infallibility, very much in line with Paul's teaching and
Thomism, but obviously not with Whitehead's God, devoid
of the faculty of omniscience and open to the irruption of the
unexpected. On the other hand, his criticisms of the Protestant
churches were no less severe: 'Calvin and Luther made the egre-
gious blunder of throwing away the whole aesthetic appeal of
the Church, which was one of its best features. You know how
bleak the Protestant services are, little for the emotions, little or
no appeal to beauty'.[3] However, despite his activism, Whitehead
abstained absolutely from writing about God and religion until
1924–1925, when he was led to recognise that his metaphysics,
after all, necessarily involved a God of some kind.

---

2   L. Price, *Dialogues of Alfred North Whitehead, As Recorded by Lucien
     Price* (New York: The New American Library, 1954), 111, 155, 247.
3   Ibid., 192.

AFTER ERIC'S DEATH

The death of Eric, killed in action on his plane on 13 March 1918 at the age of 19, his father's immense grief and, perhaps even more so, his mother's heart-wrenching grief, seem to have secretly shaken Whitehead's agnosticism. The dedication of his book *An Enquiry Concerning the Principles of Natural Knowledge*, published the following year, to his son, bears the sibylline phrase: *giving himself that the city of his vision may not perish*. It can be interpreted as transposing his own desire to be able to believe in the objective immortality of the past, and a probable source of his adoption, a few years later, of the Bergsonian theory of the persistence of the past in the present, or more precisely in his own thought, of the prehension of a large number of past events in every present occasion, thus giving human immortality a kind of objectivity. When, in 1926, he published *Religion in the Making*, Whitehead returned to the example of maternal love and gives us to think, with Lowe, that the concrete experience of the depth of Evelyn's feelings for her favourite son, as well as her grief at the announcement of his death, made the trumpets of the evidence of religious value burst inside him, evidence so convincing 'that it suffers no appeal'.[4] In any case, while Bertrand Russell's abrupt remark that 'Eric's death made him want to believe in immortality' may seem excessive, his daughter Jessie did reveal to Lowe that she thought Eric's death was surely something to do with his return to religiosity.[5]

Alex Parmentier thinks, on reading Whitehead, that he saw that the religious attitude was a fundamental dimension of

---

4   V. Lowe, *Alfred North Whitehead, the Man and His Work, Volume II: 1910-1947* (Baltimore: The John Hopkins University Press, 1990), 196.
5   Ibid., 188.

man, and that he understood that this attitude 'was not extrinsic to metaphysics, and that not only it could, but it *should*
be a cooperation between religious intuition and speculative
intelligence'.[6]

In *Science and the Modern World* published in 1925,
Whitehead set out a nuanced position towards religion. Thus,
he writes,

> [Religion] is the vision of something which stands beyond,
> behind, and within, the passing flux of immediate things;
> something which is real, and yet waiting to be realised; some-
> thing which is a remote possibility, and yet the greatest of
> present facts; something that gives meaning to all that passes,
> and yet eludes apprehension; something whose possession is
> the final good, and yet is beyond all reach; something which
> is the ultimate ideal, and the hopeless quest.

This attitude, which sees religion as a supreme good, still
shrouded in mystery, towards which one tends, and not a relationship already given, resonates well with the approach to
religion by Samuel Alexander, his colleague in the Aristotelian
Society since his London years.[7]

READING *SPACE, TIME, AND DEITY*

Around May 1924, Whitehead decided to read *Space, Time,
and Deity*. He adds many annotations in the margins of this
book, showing his interest. On the other hand, the last three
chapters, on Deity, are not annotated. One may think, however,

---

6   A. Parmentier, *La Philosophie de Whitehead et le problème de Dieu*
     (Paris, France: Beauchesnes, 1968), 537.
7   A. N. Whitehead, *Science and the Modern World* (New York:
     Macmillan, 1925), 191; S. Alexander, *Space, Time and Deity* (London:
     Macmillan, 1920).

that this is not an indication that he skipped this reading, but rather that his own position on the problems discussed was not yet fixed at that time. In the end, despite obvious differences in approach to the status and role of God in the world, in Whitehead and Alexander we find a great convergence in an eschatological understanding of the world's becoming: far from an all-powerful creator of a given once and for all, their God works for the completion of creation and its harmonisation. Rather than the eternal being out of the world, his immanence is affirmed and his own transformation, its own enrichment, invoked. As far as we can consider him as a person, we could say that he reaches out his arms to us, but lets the living forces act in the world and work its dough.

Beyond these convergences, there are also profound differences between the two thinkers. For Alexander, it is necessary to distinguish between the actual God and the quality of Deity, which is the one he aspires to and the one he works for.

> As actual, God does not possess the quality of deity but is the universe as tending to that quality. This *nisus* in the universe, though not present to sense, is yet present to reflection upon experience. Only in this sense of straining towards deity can there be an infinite actual God. [...] There is no actual infinite being with the quality of deity; but there is an actual infinite, the whole universe, with a nisus to deity; and this is the God of the religious consciousness, though that consciousness habitually forecasts the divinity of its object in an individual form. [...] Thus, if this interpretation is correct, the object of religious sentiment is no mere imagination which corresponds to a subjective and possibly illusory movement of mind. We are in perpetual presence of this object, which stimulates us, some of us more, some less; is

sometimes felt and sometimes left unexperienced according to our condition.[8]

These thoughts are summarised a few pages later in these terms: 'Deity is some quality not realized but in process of realisation, is future and not in present'.[9] Whitehead accepted the idea of a reciprocal transformation of the humanity, living beings, and things, on the one hand, and his God (the principle of concrescence and creative evolution), on the other hand, but he rejected the dichotomy between God and the principle of deity. Whitehead's God is all one, it is the dynamic principle of concrescence and creative evolution, an ever-present and ever-active actual entity, though without materiality. Dorothy Emmet has noticed that the God of Whitehead is finally closer to the God of Leibniz than to the God of Spinoza, while the reverse is true of Alexander.[10] For the God of Leibniz created the best of all possible worlds, and—like Whitehead's—continues to work for an even better world. Alexander's God appears more personal, more dualistic too (having a body—the whole universe, or more precisely space–time—and a distinct spirit, deity, or time raised to transcendence).

WHITEHEAD'S GOD AND TIME

For a long time, Whitehead had been obsessed with the complexity, not to say the magic character or majesty of time. In a

---

8   Alexander, *Space, Time and Deity*, 361–77. Alexander speaks of the infinity of God because he assimilates his body, not to the actual entities like Whitehead would tend to do so, but to the mathematical Space–Time, to which it does not assign a limit.
9   Alexander, *Space, Time and Deity*, 379.
10   D. Emmet, "Whitehead and Alexander," *Process Studies* 21 (1992): 137–48.

letter he wrote to Lord Haldane, undated but probably written in September or October 1921, Whitehead confided: 'The
problem of time seems very fundamental. In some respects *it
extends beyond nature.*[11] It expresses the fundamental activity of existence. This activity governs the general character of
nature, which is an abstraction from the concrete reality of fact.
Thus time in nature is a special aspect of the grand problem of
philosophy'.[12] It can be concluded that for him, from this date
onwards, the complexity of the problem of time was an index
of transcendence, or rather, that it opens the door to a certain
form of spirituality. Soon he was to confront these ideas with
those of the author of *Space, Time and Deity*, who for his part
had written: 'The question whether God is in Time or out of
it has been answered explicitly, and is answered implicitly by
the whole tenor of this inquiry. God's body is not spaceless nor
timeless, for it is Space-Time itself. His deity is located in an
infinite portion of Space-Time, and it is in fact essentially in
process and caught in the general movement of Time. [...] But
no theoretic consideration sustains the belief in a God who
precedes his universe'.[13] An analysis that emboldened him to
write:

> Time is an element in the stuff of which the universe and all
> its parts are made, and has no special relation to mind, which
> is but the last complexity of Time that is known to us in finite
> existence. Bare Time in our hypothesis, whose verification

---

11 I have italicised this passage in order to favour a possible
    interpretation.
12 Letter from Whitehead to Haldane, 1921, personal communication
    from Ronny Desmet, who notices that the continuation of this letter
    contains the first clear indication for the atomic nature of time: not
    only does the instant not exist, but concrete time is given in packets.
13 Alexander, *Space, Time and Deity*, 399, 421.

has been in progress through each stage of the two preceding Books and will be completed by the conception of God— bare Time is the soul of its Space, or performs towards it the office of soul to its equivalent body or brain.[14]

Many more or less materialistic readers will mock and strongly reproach its author for these views. But did not Bergson write that the law of growth of entropy (time in its dimension of agent of randomness) 'was the most metaphysical of the laws of physics'[15]? For Whitehead too, time is the most spiritual of the concepts invented by man in the face of evidence of the complexity of the process. It is the close link between time and process that gives time its spiritual character. Far from considering God first of all as an eternal being in the sense of a being outside of time, who revealed himself to men by saying 'I am who I am', according to Whitehead we must consider God above all as the motor of concrescence, the source of time and creativity, and 'he who comes'. As will be recalled a little later, Whitehead, however, would qualify this affirmation, because for him God also retains a timeless aspect under his condition of God-primordial. The fact remains that this God does not dictate, does not decree, he calls: 'God, in his primordial nature, is unmoved by love for this particular, or that particular, for in this foundational process of creativity, there are no preconstituted particulars. In the foundations of his being, God is indifferent alike to preservation and to novelty. He cares not whether an immediate occasion be old or new, so far as concerns derivation from its ancestry. His aim for it is depth of satisfaction as an intermediate step towards the

---

14 Ibid., 345.
15 H. Bergson, *L'Evolution créatrice* (Paris, France: Alcan, 1907); *Creative Evolution* (trad. A. Mitchell, Harvard: The Gutenberg Ebook Project, 2008), 243.

fulfilment of his own being. His tenderness is directed towards each actual occasion, as it arises'.[16]

GOD AS THE DRIVING FORCE FOR CONCRESCENCE AND CREATIVITY

Let us now see why, for Whitehead, God is not an option but a necessary agent for the operability of his metaphysics. A short article that William E. Hocking, one of his students at Harvard, devoted to "Whitehead as I Knew Him", reports: 'Of the concept of God, primordial and consequent,[17] he said to me: *I should never have included it, if it had not been strictly required for descriptive completeness. You must set all your essentials into the foundation. It's no use putting up a set of terms, and then remarking, "Oh, by the by, I believe there's a God"*'.[18] In *Science and the Modern World*, Whitehead explains why: 'In the place of Aristotle's God as Prime Mover, we require God as the Principle of Concretion'.[19] In reality, this new attribution may appear as a modern reformulation of the ancient formulation. God is the one who controls the mechanism of the process, making the potential entities pass from the chaos of potentiality to the concreteness of the actual entity through concrescence, with its own temporal thickness, before bringing about its perishing, until this binary cycle repeats itself again and again. In *Process and Reality*, he clarifies what he means by the 'principle of concretion': it is 'the principle whereby there is initiated a definite outcome from a situation otherwise riddled with ambiguity'.[20]

---

16 A. N. Whitehead, *Process and Reality* (New York: Macmillan, 1929), 105.
17 These divine attributes will be defined in the following paragraph.
18 W. E. Hocking, "Whitehead as I Knew Him," *The Journal of Philosophy* LVIII (1961): 505–16.
19 Whitehead, *Science and the Modern World*, 174.
20 Whitehead, *Process and Reality*, 345.

We recognise here the mechanism of concrescence; it reminds us of the collapse of the wave function in quantum mechanics, but Whitehead refuses to explain it by appealing to a magical power of the observer's mind, as many of the founders of quantum mechanics did.

Whitehead's reference to Aristotle may be surprising, when one knows his preference for Plato. But in this circumstance, it is natural. This is because, as Alex Parmentier expresses it, 'If Aristotle's primary philosophy necessarily implies the existence of an "eternal being, substance and pure act",[21] Whitehead's philosophy, we believe, cannot be divorced from his God either. And this God—on which we will reflect further—we do not think that we can simply reduce him to a logical schema. For we are convinced that Whitehead was a sincerely religious thinker and, in the words of Ch. Hartshorne,[22] *a very honest writer ... When he said "love" he did not mean something quite different, such as the "scheme of extension" with which Mays[23] apparently identifies Whitehead's God'.*[24]

And Madame Parmentier further insists on the idea that it is above all to Aristotle's principles that one must confront his philosophy: 'If Whitehead once said, as a sort of joke, in speaking of the "savage hostility" of the Church towards Galileo, that it was the characteristic of men whose erudition surpasses genius, to start quoting Aristotle when they were shocked, he nevertheless invited us to confront him with Aristotle by taking

---

21 Aristotle, ca. 322 bc. Metaphysics, A7, 1072a25.
22 Charles Hartshorne (1897–2000) is an American philosopher, specialist in metaphysics and philosophy of religions, deeply influenced by Whitehead. He was the main architect of the development of the Process Theology after the Second World War.
23 Wolfe Mays (1912–2005) is an English philosopher, specialist in phenomenology. He had a critical reading of Whitehead.
24 Parmentier, *La Philosophie de Whitehead*, 536.

a position in relation to him, in an explicit way, on almost all the important points of his metaphysics'.[25]

In *Religion in the Making*, Whitehead returns to the justification of the recourse to God. He writes:

> The order of the world is no accident. There is nothing actual which could be actual without some measure of order. The religious insight is the grasp of this truth: That the order of the world, the depth of reality of the world, the value of the world in its whole and its parts, the beauty of the world, the zest of life, the peace of life, and the mastery of evil, are all bound together—not accidentally, but by reason of this truth: that the universe exhibits a creativity with infinite freedom, and a realm of forms with infinite possibilities; but that this creativity and these forms are together impotent to achieve actuality apart from the completed harmony, which is God.[26]

Clearly, if Whitehead needs a God, it is first of all because of the order that he discovers in the world, this two-tone music that repeats itself tirelessly: the concrescence and the perishing of each actual entity. God is the agent of concrescence. By this very fact, he is also the motor of creativity, the spirit that imbues each actual entity, makes it want to insert a dose of novelty in each act of concrescence. He accomplishes this creativity in two ways: he inhabits the mental pole of each actual entity in the process of becoming, by 'suggesting' the sorting of the actual entities of the past that he wants, or not, to re-actualise through the mechanism of prehension in the current entity that is about to become by a new concrescence. But he is also the guardian of an infinity of as yet unrealised entities, conceptual

---

25 Ibid., 574.
26 A. N. Whitehead, *Religion in the Making* (New York: Macmillan, 1926), 107.

and potential, which Whitehead calls the eternal entities, and which we discussed in Chapter 6.

However, Whitehead wishes to warn us against the image of a creator God, pre-existing to the world, as he is presented in traditional religions. In a sense, writes Whitehead, 'God can be termed the creator of each temporal actual entity. But the phrase is apt to be misleading by its suggestion that the ultimate creativity of the universe is to be ascribed to God's volition. The true metaphysical position is that God is the aboriginal instance of this creativity, and is therefore the aboriginal condition which qualifies its action. It is the function of actuality to characterize the creativity, and God is the eternal primordial character. But, of course, there is no meaning to "creativity" apart of its "creatures", and no meaning to "God" apart from the "creativity" and the "temporal creatures", and no meaning to the "temporal creatures" apart from "creativity" and "God".[27] David Ray Griffin, Professor of Philosophy and Religion at the University of Claremont, develops this idea and comments:

> God is not only creativity; God has determinate characteristics: God knows the world, envisages primordial potentials with appetition and purpose, influences the world, and is in turn influenced by the world. [...] God is the primordial embodiment of Creativity. (Creativity is ... not an actuality that could exist by itself, unembodied). Whitehead refers to God as the "eternal primordial character" of Creativity.[28]

On reading the foregoing, one might be tempted to accuse Whitehead of having a uniformly progressive view of evolution,

---

27 Whitehead, *Process and Reality*, 225.
28 D. R. Griffin, "Bohm and Whitehead on Wholeness, Freedom, Causality, and Time," in *Physics and the Ultimate Significance of Time*, ed. D. R. Griffin (Albany: State University of New York, 1986), 137.

in line with the orthogenesis popular at the time Whitehead wrote his treatises. However, Whitehead hardly ventured into this terrain in his books. In fact, it is clear that he did not adhere to orthogenesis. That is why he distinguished between God and the world: God suggests, but it is the actual entities that decide the form of their concrescence: 'The divine element in the world is to be conceived as a persuasive agency and not as a coercive agent'.[29]

PRIMORDIAL AND CONSEQUENT NATURE OF GOD

For we must now, with Whitehead, distinguish in God two contrasting attributes. First, there is the primordial God. As we have just seen, this is the principle of order, and it is the motor of concrescence, the one who gives the latter a touch of new-ness to the concretised actual occasions, and the one who holds the key to all the eternal entities and can make them intervene at each concrescence. The eternal entities are like the stars in his mantle. Under this attribute, God is primordial, not in the sense of an antecedent to creation, but in the sense of an immutable energy, foundational of the universe and of its evolution. When Whitehead wrote, in 1925, *Science and the Modern World*, he might sometimes suggest that he is still hesitating to give this omnipresent and omni-active energy the name of God; thus, in the following paragraph: 'What is the status of the enduring stability of the order of nature? There is the summary answer, which refers nature to some greater reality standing behind it. This reality occurs in the history of thought under many names, The Absolute, Brahma, The Order of Heaven, God. The

---

29 A. N. Whitehead, *Adventures of Ideas* (New York: Free Press, 1933), 166. In this passage, Whitehead specifies that he borrows this sentence from Plato.

delineation of final metaphysical truth is no part of this lecture. My point is that any summary conclusion jumping from our conviction of the existence of such an order of nature to the easy assumption that there is an ultimate reality which, in some unexplained way, is to be appealed to for the removal of perplexity, constitutes the great refusal of rationality to assert its rights. We have to search whether nature does not in its very being show itself as self-explanatory.'[30]

But it is a pedagogical skilfulness, because he himself has already answered this perplexity in the negative. In the final chapter of the same book, and then more clearly in *Process and Reality*, he clarifies his thinking by distinguishing between philosophical necessity and religious intuition: 'The concept of God is certainly one essential element in religious feeling. But the converse is not true; the concept of religious feeling is not an essential element in the concept of God's function in the universe. In this respect religious literature has been sadly misleading to philosophical theory, partly by attraction and partly by repulsion.'[31] Nevertheless, the fact remains that by reading *Process and Reality*, we see that Whitehead does not hesitate to reaffirm forcefully the presence of a God and his action in creativity: 'The primordial nature of God is the acquisition by creativity of a primordial character.'[32]

The God-Primordial is the centre of a creative and teleological power. 'Creative', first, 'not in the sense of a creation *ex nihilo*, but in the sense that the self-creation of the present entities has its source in God, and without him could not be realized'. 'Creative' again in the sense that this evaluation of the possible by God is not conditioned by anything: ' "its unit of

---

30 Whitehead, *Science and the Modern World*, 94.
31 Whitehead, *Process and Reality*, 207.
32 Ibid., 344.

conceptual operations is a free creative act (*Process and Reality* 522)"—for in its primordial nature God presupposes nothing, no determination already given; he presupposes only creativity'. Teleological, on the other hand, because 'his conception of potentiality implies an *appetite*, that is, an aspiration to realise what he conceives, with a view to his own full realisation— through the full realisation of the *creatures*'.[33]

But the attribute of God's primordiality does not exhaust its nature. Motor of concrescence and creativity, its action is quite physical and automatic: to make the concrete reality happen from potentiality, to breathe into it, when the conditions are convenient, and when actual occasions in the process of concrescence may contribute to it, a dose of novelty. Until then, his God could be Spinoza's *Deus sive natura*. But Whitehead is not satisfied with this divine dimension. He needs a dynamic, spiritual, and relational dimension in relation to the world, because for him God and the world are mutually transforming each other. And this constructive dialogue between the world and himself defines another facet of Whitehead's God, his consequent nature. Whitehead defines it in *Process and Reality*:

> His 'consequent nature' results from his physical prehension of the derivative actual entities. [...] By reason of its character as a creature, always in concrescence and never in the past, it receives a reaction from the world; this reaction is its consequent nature. It is here termed 'God'; because the contemplation of our natures, as enjoying real feelings derived from the timeless source of all order, acquires that 'subjective form' of refreshment and companionship at which religions aim.[34]

---

33 Parmentier, *La Philosophie de Whitehead*, 548.
34 Whitehead, *Process and Reality*, 31.

'The consequent nature of God is the progressive realisation of his full (and no longer merely conceptual) actuality, by means of his assimilation (his "prehension") of the actual entities of the temporal world. It is the dynamic, evolutionary and self-creating aspect of God. God appropriates the world and transforms it into his own being, transfiguring it, and in this way he creates himself incessantly'.[35] This leads us to Whitehead's statement in *Process and Reality*: 'The consequent nature of God is the fluent world become 'everlasting' by its objective immortality in God', a phrase he himself explains in an already mystical thrust: 'The consequent nature of God is his judgment on the world. He saves the world as it passes into the immediacy of his own life. It is the judgment of a tenderness which loses nothing that can be saved. It is also the judgment of a wisdom which uses what in the temporal world is mere wreckage. [...] God's role is not the combat of productive force with productive force, of destructive force with destructive force; it lies in the patient operation of the overpowering rationality of his conceptual harmonization. He does not create the world, he saves it: or, more accurately, he is the poet of the world, with tender patience leading it by his vision of truth, beauty, and goodness'.[36]

RELIGION IN THE MAKING

In publishing the Lowell Lectures of 1925 in *Science and the Modern World*, Whitehead came to the decision to add a final chapter on God to the simple transcript of his lectures; he wanted to seize the opportunity to introduce and explain his reasons for using this entity in his metaphysical system. 'In *Science and the Modern World*', writes Alex Parmentier, 'after having

---

35 Parmentier, *La Philosophie de Whitehead*, 548.
36 Whitehead, *Process and Reality*, 346–47.

shown the necessity of positing a Principle of Concretion that is God, Whitehead defines what he considers to be the essence of the religious spirit—which corresponds for him, let us recall, to "integral experience". The religious spirit is one that perceives a permanence beyond, and within, the flow of immediate things; it is one that perceives an order disposing all things in love (and not in force) for the achievement of eternal harmony; and the immediate reaction of human nature to this vision is worship, a worship of love tending towards assimilation into a reciprocal love".[37]

But the final chapter of *Science and the Modern World* is brief and leaves many questions open. The Lowell Institute therefore decided to offer Whitehead the possibility of a new series of four lectures, this time at King's Chapel in Boston, to present his views on religion and theology in greater detail. Lowe reports that in July 1925 Whitehead broke the news to his son North, writing to him, 'I am going to give a new series of Lowell lectures next year, on the subject of "Science and Religion", that is, on the scientific criticism of religion'. One would think that he is preparing for a diatribe against the irrationality of religious feeling. But in the end, it is the best presentation of his views on God, the place he occupies in his metaphysics, his convergences and divergences from the theology of the main religions—Buddhism and Christianity in particular—that he delivers in *Religion in the Making*. The tone is measured, calm, far from the pamphlet that his remark to his son North foreshadowed. Whitehead seems rather delighted to put on paper, in a concise but systematic way, his approach to the role of God in the world, and what he sees as the future of the world's religions.

---

37 Parmentier, *La Philosophie de Whitehead*, 548.

The first thing that strikes one when reading this book is the height of vision it adopts from the beginning. It highlights how religions, from the rituals of primitive peoples to the great religions, have helped organise society and advance socially. For Whitehead, confronting what they have in common reveals profound truths, useful for the progress of humanity; confronting their differences allows us to uncover relative or unfounded beliefs and arbitrary practices. This is particularly the case with belief in a personal God: 'Christian theology has also, in the main, adopted the position that there is no direct intuition of such an ultimate personal substratum for the world. It maintains the doctrine of the existence of a personal God as a truth, but holds that our belief in it is based upon inference. Most theologians hold that this inference is sufficiently obvious to be made by all men upon the basis of their individual personal experience. But, be this as it may, it is an inference and not a direct intuition. [...] The wisdom of the main stream of Christian theology in refusing to countenance the notion of a direct vision of a personal God is manifest. For there is no consensus'.[38]

Thus, according to Whitehead, religions (including primitive religions) have, on the whole, promoted the organisation of society and social progress. But this is no reason to hide their bad sides: 'Indeed history, down to the present day, is a melancholy record of the horrors which can attend religion: [...] The uncritical association of religion with goodness is directly negatived by plain facts. Religion can be, and has been, the main instrument for progress. But if we survey the whole race, we must pronounce that generally it has not been so: *Many are called, but few are chosen*'.[39]

---

38 Whitehead, *Religion in the Making*, 53.
39 Ibid., 45.

So far Whitehead has not considered the concrete example of the Christian religion and the life of Christ. He does so in the second lecture, emphasising the adequacy of this life to the propensity for action of the God that he believes he sees at work in the world: 'The life of Christ is not an exhibition of over-ruling power. Its glory is for those who can discern it, and not for the world. Its power lies in its absence of force. It has the decisiveness of a supreme ideal, and that is why the history of the world divides at this point of time'.

These free words end with a historical assessment: 'There are three main simple renderings of [the concept of God] before the world:

1. The Eastern Asiatic concept of an impersonal order to which the world conforms. This order is the self-ordering of the world; it is not the world obeying an imposed rule. The concept expresses the extreme doctrine of immanence.

2. The Semitic concept of a definite personal individual entity, whose existence is the one ultimate metaphysical fact, absolute and underivative, and who decreed and ordered the derivative existence which we call the actual world. This Semitic concept is the rationalization of the tribal gods of the earlier communal religions. It expresses the extreme doctrine of transcendence. 'But this approach to God raises two difficulties. It leaves God completely outside a metaphysical rationalisation, and does not respond to the difficulty of proving his existence. The only possible proof appears the one imagined by Anselm, the 'ontological proof'. According to this proof, the mere concept of such an entity allows us to infer its existence. Most philosophers and theologians reject this proof: for example, it is explicitly

rejected by Cardinal Mercier in his *Manual of Scholastic Philosophy*'.[40]

3. The pantheistic concept obviously is close to Spinoza's God. According to this approach, the present world, conceived as distinct from God, is not real. Its only reality is that of God. This is the extreme doctrine of monism. For Whitehead, the Semitic concept of God can very easily be transformed into a pantheistic concept: 'In fact, the history of philosophical theology in various Mahometan countries such as Persia, for instance shows that this passage has often been effected'.

In the third lecture, Whitehead begins by recalling his choice for a non-personal God: 'If, at this stage of thought, we include points of radical divergence between the main streams, the whole evidential force is indefinitely weakened. Thus religious experience cannot be taken as contributing to metaphysics any direct evidence for a personal God in any sense transcendent or creative'. But at the same time, he insists once again in his plea for the existence of a God, in line with Kant's arguments: [My] line of thought extends Kant's argument [for the existence of God]. He saw the necessity for God in the moral order. But with his metaphysics he rejected the argument from the cosmos. The metaphysical doctrine, here expounded, finds the foundations of the world in the aesthetic experience, rather than as with Kant in the cognitive and conceptive experience. All order is therefore aesthetic order, and the moral order is merely certain aspects of aesthetic order. The actual world is the outcome of the aesthetic order, and the aesthetic order is derived from the immanence of God. [...] According to the doctrine

---

40 Ibid., 58.

of this lecture, every entity is in its essence social and requires the society in order to exist. In fact, the society for each entity, actual or ideal, is the all-inclusive universe, including its ideal forms'.[41]

But Whitehead insists at the same time on the fact that divine action does not extinguish but on the contrary magnifies the freedom of the present entities, even if it offers them orientations. Amongst the factors shaping the universe: creativity, eternal entities, one must add 'the present but non-temporal entity by which the indeterminacy of simple creativity is transmuted into a determined freedom. This present but non-temporal entity is what men call God—the supreme God of religions made rational. [...] The actual but non-temporal entity whereby the indetermination of mere creativity is transmuted into a determinate freedom. This non-temporal actual entity is what men call God—the supreme God of rationalized religion. [...] An epochal occasion is a concretion. It is a mode in which diverse elements come together into a real unity. Apart from that concretion, these elements stand in mutual isolation. Thus an actual entity is the outcome of a creative synthesis, individual and passing'.[42] Thus, Whitehead's metaphysics tries to reconcile, for each individual entity (and not only for man), determination linked to the past and freedom. This is why he needs a God who is so subtle, the motor of concrescence but not constraining.

A few lines later, Whitehead summarises the relationship between God and the universe in creation: 'The purpose of God is the attainment of value in the temporal world. An active purpose is the adjustment of the present for the sake of adjustment of value in the future, immediately or remotely. [...] On

---

41 Ibid., 91.
42 Ibid., 80.

the other side, the occasion is the creature. This creature is that one emergent fact. This fact is the self-value of the creative act. But there are not two actual entities, the creativity and the creature. There is only one entity which is the self-creating creature. [...] The actuality is the enjoyment, and this enjoyment is the experiencing of value. For an epochal occasion is a microcosm inclusive of the whole universe. This unification of the universe, whereby its various elements are combined into aspects of each other, is an atomic unit within the real world'.[43] In these lines appears clearly in Whitehead the disciple of Leibniz, but a social and organic Leibniz.

The fourth lecture goes into the details of the organisation of religions and is interested first of all in their dogmas and their rational necessity, but also in the dangers of abusive belief in their immutability and of fundamentalism: 'A dogma is the precise enunciation of a general truth, divested so far as possible from particular exemplification. [...] A dogma—in the sense of a precise statement—can never be final; it can only be adequate in its adjustment of certain abstract concepts. But the estimate of the status of these concepts remains for determination. [...] If the same dogma be used intolerantly so as to check the employment of other modes of analyzing the subject matter, then, for all its truth, it will be doing the work of a falsehood. [...] Thus religion is primarily individual, and the dogmas of religion are clarifying modes of external expression. The intolerant use of religious dogmas has practically destroyed their unity for a great, if not the greater part, of the civilized world'.[44]

A few pages later, Whitehead continues: 'Religions commit suicide when they find their inspiration in their dogmas'.

---

43 Ibid., 88–89.
44 Ibid., 122.

Specifying his thought, he explains that if one can admit the
necessity of dogmas to give coherence to a religion, one must
note that historically each religion has suffered from lock-
ing itself into a fixed interpretation of its dogmas, not leaving
enough room for their co-evolution with society. He believes
that the recent decline of Christianity and Buddhism, in particular, is partly due to the fact that each religion has unduly
barricaded itself to the other: 'The self-sufficient pedantry of
learning and the confidence of ignorant zealots have combined
to shut up each religion in its own forms of thought. Instead of
looking to each other for deeper meanings, they have remained
self-satisfied and unfertilized. Both have suffered from the rise
of the third tradition, which is science, because neither of them
had retained the requisite flexibility of adaptation.'[45]

In the book conclusion, Whitehead returns to God's role in
the world in his philosophy: 'God is that function in the world
by reason of which our purposes are directed to ends which in
our own consciousness are impartial as to our own interests.
He is that element in life in virtue of which judgement stretches
beyond facts of existence to values of existence. He is that element in virtue of which our purposes extend beyond values
for ourselves to values for others. He is that element in virtue
of which the attainment of such a value for others transforms
itself into value for ourselves. He is the binding element in the
world. The consciousness which is individual in us, is universal
in him: the love which is partial in us is all-embracing in him.
Apart from him there could be no world, because there could
be no adjustment of individuality.'[46]

---

45 Ibid., 131.
46 Ibid., 143.

MYSTICISM AND PROCESS THEOLOGY

After the publication of *Religion in the Making* and then *Process and Reality*, Whitehead seems to have gradually accentuated his mystical vision of the world, his sense of solidarity, and God's call to universal love. But John Cobb, as a Christian, warns us: 'Whitehead did not regard himself as a theologian or even, necessarily, as a Christian. Indeed, at times he spoke harshly of both theology and the church. Still it is far from artificial to claim him, too, in his envisioning of hope, as a servant of Christ'. For, Cobb says, 'Whitehead's thought of God is centred on the Logos', that energy, that Johannine verb, that principle of order, 'which he called the Primordial Nature'.[47]

Alex Parmentier continues along the same lines: 'It is indisputable that Whitehead is a religious thinker. That his God is not the God of Christian Revelation (although it will be advisable to qualify this judgment), that his conception of religion and his metaphysics have an aesthetic flavour that brings them singularly close to Pythagorism, we certainly agree and we will come back to this. Be that as it may, Whitehead is a religious thinker, and three reasons seem to us sufficient to define him as such: For him the universe could not exist without God (even if God does not create it ex nihilo); without God existing reality is unintelligible. He saw that the essence of the religious attitude was adoration. Obviously this adoration, in Whitehead's case, is not addressed to a Creator in the strong sense of the term; the fact remains that for Whitehead, man could not exist without God, and that the recognition of his relation to God is translated into adoration, and adoration of love'.[48]

47　J. Cobb, *Christ in a Pluralistic Age* (Philadelphia: Westminster Press, 1975), 340, 335.

48　Parmentier, *La Philosophie de Whitehead*, 537.

Whitehead's deistic and mystical cosmology reminds us of Teilhard de Chardin; this proximity has often been noted, especially because of the idea of continuous creation, manifested by evolution. Teilhard's interpretation of evolution as continuous creation meets Whitehead's thinking about God's role in the world quite well, and Teilhard's imagined final end as the unification of collective consciousness with the spirit of God in the Omega point resonates quite well with the eschatological end that Whitehead proposes. However, one should not hide the profound differences between the two thinkers. Not only because Whitehead rejects the idea of an original creator God, and that there is no Alpha point for him, but a big-bang and Platonic chaos; more important is their difference in their views of God's role in the world. Even if Christ in Teilhard is not only the historical figure of the incarnate God but the pledge of a universal God, of an immanent 'logos', Teilhard's primordial God remains the traditional God of the Church, he has a definite project for the world, that of a unification of the world in his love. In contrast to Whitehead's God, Teilhard's God—like that of the Christian churches—does not really change, he is content to imprint a definite direction and to assign a glorious final end to the entire universe, with the help of the collective conscience of humanity.[49] Whitehead, on the other hand, is much less teleological. God is the creator of novelty, he suggests an ideal of harmony, but it is the actual entities that choose in

---

49 It should be noted, however, that in *The Heart of Matter*, Teilhard suggests that in the relations between God and the world, there is a certain reciprocity and therefore a certain common evolution: 'In the vast phenomenon of Christification that the conjunction of the World and God discovered for me [...] I am led to see now (in accordance with the spirit of Saint Paul) a mysterious product of completion and fulfilment for the Absolute Being himself' (P. Teilhard de Chardin, *Le cœur de la Matière* (Paris, France: Le Seuil, 1976), 64).

their concrescences their way of incarnating it. This entails a more diverse world, and a more chaotic path towards the ideal and the ultimate end.[50]

Whitehead's insistence on the dynamic reciprocity between God and the world inspired some striking aphorisms, which became famous amongst metaphysical philosophers: 'Either of them, God and the World, is the instrument of novelty for the other'. 'It is as true to say that God is permanent and the World fluent, as that the World is permanent and God is fluent'. 'It is as true to say that the World is immanent in God, as that God is immanent in the World'. 'It is as true to say that God creates the World, as that the World creates God'.[51]

At the end of his thesis, Alex Parmentier wonders: 'The intuition of love—and of God, since God is love—cannot, therefore, operate in Whitehead the last synthesis; it must necessarily allow itself to be reduced to that of creativity. But how can this be explained when one has grasped the depth of Whitehead's Johannine intuition of God-Love? Is there a mystical Whitehead, for whom Love is first, and a philosopher Whitehead, for whom creativity comes first? [...] Let's not forget that, metaphysically, within Whitehead's system, God is not a person, but an actual entity [...] Whitehead's metaphysics, where God is not, in the strictest terms, a person, cannot be a metaphysics of love, and leads to a kind of immanentism; love is not that of the mystic who is "lost" in a Person; it is that of the

---

50 See on this respect G. Barbour, "Teihard's Process Metaphysics," *Journal of Religion* 49 (1969): 136–59; J. S. Homlish, *The Cosmos and God according to Pierre Teilhard de Chardin and Alfred North Whitehead* (Philosophical diss., University McMaster, 1974), accessed November 24, 2021, https://macsphere.mcmaster.ca/handle/11375/15525.

51 Whitehead, *Process and Reality*, 348–49.

drop of water that is lost in an ocean that is not a Person, but the Whole: God-and-the-World'.[52]

Whitehead's religious thought can be assumed as complex; it is a narrow track, an uneasy ridge path. It is the option of panentheism, between the totally immanent God of Spinoza and the personal God of the revealed religions. Those who are tempted to follow him today in his mystical approach and are fully aware of the contributions of contemporary science must beware of the temptation of Spinoza's pantheism, which according to Whitehead is incapable of accounting for the creativity of the process, while maintaining a critical spirit towards the instituted religions. For him, these are never more than sketches, even if they have shed light on some of the mysteries of the organisation of the world.

---

52 Parmentier, *La Philosophie de Whitehead*, 568.

Epilogue

After the death of Alfred North Whitehead, who died of a cerebral haemorrhage on 30 December 1947, the fame of this thinker began to grow, mainly in the United States. Charles Hartshorne (1897–2000), who had worked with Whitehead for some time, set about the task of building bridges between Whitehead's metaphysics and Protestant Christianity. After all, we must examine what prevents us from personalising Whitehead's God, without transforming him into that Michelangelo's God of the Sistine Chapel, who would have pre-existed the world and created Man with a sovereign finger, and from seeing in Jesus the Christ, that is to say, the initiate par excellence of God, and as John Cobb[1] says after the Evangelist, the God of the Logos, the Incarnate Word. What prevents religions from amplifying the timid movements of dialogue that have emerged in recent decades, in order to reach a reasonable hope that, in the end, thoughts as different as Christianity and Buddhism will find more and more points of convergence? John Cobb took up the torch, notably with David Ray Griffin. In 1973, they founded and animated, and still do today, the *Center for Process Studies* in Claremont, near Los Angeles. Numerous studies have been published as part of the Center's activities. Let us mention on this subject the book by John Cobb and David Ray Griffin, *Process Theology: An Introductory Exposition*, 1976 (Philadelphia: The Westminster Press), and *Christ in a Pluralistic Age*, which we have already

---

1   J. Cobb, *Christ in a Pluralistic Age* (Philadelphia: Westminster Press, 1975), 335.

met. In France, the penetration of Whitehead's philosophy came later. In process theology, we should mention *Le Dynamisme Créateur de Dieu*, by André Gounelle, 2000 (Paris: Van Dieren). Although the centre of gravity of the thinkers and theologians it brings together belongs to the Protestant movement of Christianity, the partisans of process theology, faithful to the spirit of Whitehead, believe that each religion, including Buddhism, must eventually makes its contribution and that all theologies should converge towards a final theology, reformed by adapting dogmas to scientific advances and social progress; if we look at the history of religions in an objective way, we can only observe that this is already the case. In the Catholic religion, for example, the official explanation of dogmas is no longer quite the same as it was taught during the Renaissance or at the councils that gave birth to them.

Whitehead's fame also spread amongst the philosophers of science. Today, a biennial congress brings together some six-hundred participants, epistemologists, philosophers, or scientists in the human and life sciences, all specialists in Whitehead. A minority of physicists, but not the least important ones, have so far taken an interest in Whitehead's philosophy of science; but it must be recognised that they question the difficulties of adapting this author's philosophy, not to the fundamental ideas, but to the details and current data of quantum physics and relativistic cosmology. They also expected to find, in accordance with the spirit of the philosophy of the organism, that the laws of nature and the universal constants that they include, change as they adapt to each phase of the evolution of the world; but they did not find any evidence in this sense. On the contrary, the apparent constancy of the fundamental parameters of physical theory over billions of years is impressive. Are these laws of Nature, these universal constants, therefore timeless and immutable dictates, and if so, whose dictates?

Whatever one thinks of Whitehead's metaphysics as a whole, his work must first be evaluated in terms of its main contribution: a surgical analysis of the notion of *process* and its relations with the common and scientific concept of time. Certainly, he was not the first to notice the inexistence of the instant. Thus, in a letter dated 1663 and not yet discovered during Whitehead's lifetime, Spinoza wrote: 'Wherefore many, who are not accustomed to distinguish abstractions from realities, have ventured to assert that duration is made up of instants, and so in wishing to avoid Charybdis have fallen into Scylla. It is the same thing to make up duration out of instants, as it is to make number simply by adding up noughts'.[2] Which for mathematicians is an irrefutable fact. But Whitehead has gone further than his predecessors in this area. He attempted a more methodical and rigorous analysis of the complexity of the process and showed that it was absolutely necessary to distinguish between this primary reality, the foundation of the world's fluidity but also of its irreversible advancement, and the time of clocks or physicists. Like Bergson, he understood that the present is enriched by the past, but he proposed a mechanism to explain it (that of prehensions). Like Heisenberg, he understood the importance of the distinction between potentiality and concreteness, but he proposed the mechanism of concrescence to explain the singularity of the concrete. In this, he opened a promising way to better understand the paradoxical aspects of quantum mechanics and measurement theory. Like Alexander, he saw that the affairs of the world gave rise to novelty, to concrete emergences, but he wanted to be more realistic than the latter by making spatio-temporal events, created by concrescence, the bricks of the world, rather than thinking of a concrete space–time. What Whitehead brought most valuable to the philosophy of science,

---

2   B. Spinoza, *Letter XII* (1663), addressed to Louis Meyer.

however, was undoubtedly his attempt to give an ontological status to the present. Not many philosophers have ventured to do so, and even fewer physicists. Let us recall that Whitehead introduced the word concrescence into his language in 1926, the year before Heisenberg introduced the collapse of the wave function mechanism, which concrescence inevitably brings to mind.

Before using this term, Whitehead used the term *supersession*, in the sense of 'replacement'. It is therefore the idea that concrete reality is continually replacing itself that is at the basis of the notion of process, and then of concreteness. And the present is, before 1926, the incredibly fleeting state of a global supersession. Later, in *Process and Reality*, the present becomes the moment in which, through a general concrescence, all current entities enter into coalescence, so that 'the many become one, and are increased by one', or through the transition 'from attained actuality to actuality in attainment'.[3]

Let us note that the modern philosophers who have analysed the present have all, or almost all, linked this notion to human psychology. Kant makes of time, in general, and of the present, in particular, *a priori* concepts, necessary to human understanding, even if he does not deny that something of time must also belong to nature itself. The most interesting case, with Bergson that we have already compared to Whitehead, is undoubtedly that of Heidegger. The author of *Being and Time*, who shares many intuitions close to those of Whitehead concerning time,[4] relativises the notion of the present by the tension of being towards the future, on which he insists so much. Born

---

3   A. N. Whitehead, *Process and Reality* (New York: Macmillan, 1929), 21, 214.
4   See in this respect D. Mason, "Time in Whitehead and Heidegger: Some Comparisons," *Process Studies* 5 (1975): 83–105.

of a kind of fusion or permeation between past and future, the specificity of the present becomes, in Heidegger's work, rather blurred. The originality of Whitehead's treatment of the present is that, as a scientist, he shifts the core of the question of the present from the human soul to the process of nature. Let us add that the concept of present as understood by Whitehead does not, in my opinion, betray Einsteinian relativity. In particular, the fragments of time that are born at each concrescence must naturally be understood as fragments of proper time, in the sense that the theory of relativity gives to this term.

A second valuable contribution of Whitehead's metaphysics is to open the door to a restoration of free will to humankind and other conscious beings. The negation of any reality to the instant ruins the claim to the universality of physical determinism. This door that was thought to be locked because of the physico-physiological causality of neuronal processes is probably not closed to more flexible rules of causality, expressed at different levels of the organisation of mental processes, and this restores weight to the conclusions that Roger Sperry drew from his psycho-physiological observations of commissurotomised patients. To be sure, David Griffin is right to write that 'The status of freedom, causality, and time are in the same boat. The denial of one implies, finally, the denial of the others. If time is unreal, in the sense that from an ultimate perspective there is no distinction between past, present, and future, then there can be no freedom, in the sense of self-determination in the moment. [...] Likewise, if time is unreal, then there is no real causation, as distinct from logical implication.'[5] But this only

---

5   D. Griffin, "Bohm and Whitehead on Wholeness, Freedom, Causality and Time," in *Physics and the Ultimate Significance of Time*, ed. D. R. Griffin (Albany: State University of New York, 1986), 152.

becomes completely true when we recognise the complex, 'atomic' structure of time.

Certainly, one can think that Whitehead's work is not complete. It is of course not. Like any human philosophical work, it is revisable, surpassable. Whitehead himself, with the humility that characterises him, predicted it; he, whom Victor Love remembers crossing the Harvard courtyard as a man with 'shoulders much bent, an umbrella often held across his back; his head down, but his clear blue eyes up. [...] A real face, in comparison with which others looked like mere masks'.[6] It will therefore be necessary to go beyond it. As often in the past since Galileo Galilei, science will undoubtedly be the privileged instrument to suggest the necessary philosophical, metaphysical, maybe even theological advances.

---

6   V. Lowe, *Understanding Whitehead* (Baltimore: Johns Hopkins Press, 1962), 5.

Bibliography

WHITEHEAD'S WORKS

Whitehead, A. N. "On Mathematical Concepts of the Material World." *Philosophical Transactions Royal Society London, Series A* 205 (1906), 465–525.

Whitehead, A. N. "The Axioms of Geometry." In *Encyclopedia Britannica*, 11th ed. 1910.

Whitehead, A. N. "La Théorie Relationniste de l'espace." *Revue de Métaphysique et de Morale* 23 (1916): 423–54.

Whitehead, A. N. "The Anatomy of Some Scientific Ideas." Lecture reproduced in *The Aims of Education and Other Essays*. chapter IX. New York : Macmillan, 1917/1929, 128.

Whitehead, A. N. *An Enquiry Concerning the Principles of Natural Knowledge*. Cambridge: Cambridge University Press, 1919.

Whitehead, A. N. *The Concept of Nature*. Cambridge: Cambridge University Press, 1920.

Whitehead, A. N. *The Principle of Relativity with Applications to Physical Science*. Cambridge: Cambridge University Press, 1922.

Whitehead, A. N. *Science and the Modern World*. New York: Macmillan, 1925.

Whitehead, A. N. *Religion in the Making*. New York: Macmillan, 1926.

Whitehead, A. N. *The Aims of Education and Other Essays*. New York: Macmillan, 1929/New York: Free Press, 1967.

Whitehead, A. N. *Process and Reality*. New York: Macmillan, 1929/Corrected Edition, New York: Free Press, 1979.

Whitehead, A. N. *Adventures of Ideas*. New York: Free Press, 1933/1967.

Whitehead, A. N. *Modes of Thought*. New York: Macmillan, 1938.

Whitehead, A. N. *The Interpretation of Science, Selected Essays*. Indianapolis: Bobbs-Merrill Co., 1961.

BIOGRAPHICAL TESTIMONIALS

Ford, L. S. *The Emergence of Whitehead's Metaphysics 1925–1929*. Albany: State University of New York Press, 1984.

Lowe, V. *Understanding Whitehead*. Baltimore: The John Hopkins Press, 1962.

Lowe, V. *Alfred North Whitehead, the Man and His Work, Volume I: 1861–1910*. Baltimore: The John Hopkins University Press, 1985.

Lowe, V. *Alfred North Whitehead, the Man and His Work, Volume II: 1910–1947*. Baltimore: The John Hopkins University Press, 1990.

Price, L. *Dialogues of Alfred North Whitehead, As Recorded by Lucien Price*. New York: The New American Library, 1954.

Index of Names

Glossary and Thematic Index

Distance (invariant –): distance between two events in space–time. 141, 145

Duration: in Whitehead, a limited space–time slot between two times but open in terms of space. 27, 29, 42, 46, 55, 60, 62, 65, 68, 71, 75, 79, 90, 94, 146, 153, 157, 158, 225

Emergence: the hypothesis that new properties can appear in systems that have crossed a threshold of complexity, not provided for and/or not explicable by the laws that regulate their components. 44, 182, 186–193

Entanglement: two systems that have interacted in the past cannot anymore be considered, in all rigour, as more than a single system. It is the property of non-locality, demonstrated in quantum mechanics. 109–131

Eternal entities: class of entities that do not belong to the concrete world, but are susceptible at any time to be 'prehended' by an occasion in the process of concrescence, to bring an element of novelty to the entity once concretised. 96, 101, 106, 128, 163, 164, 167, 215, 217

Evolution: 39, 41, 73, 80, 92, 100, 124, 129, 158, 163, 165–167, 169, 187, 189, 206–207, 219

Feelings: how actual occasions that are being concretised relate to other actual occasions. Physical forces are simple feelings, but for Whitehead, there may be more subtle feelings, because there is a subjectivity of actual occasions. 16, 45, 85, 92, 101–106, 153, 173, 175, 197, 208

Finality (extrinsic –): 106

Free will: the assumption that human actions can escape determinism. 169–193, 227

Future: In Whitehead, the future remains virtual. 4, 9, 35, 75, 84, 89, 90, 104–107, 124, 134, 137, 159, 200, 211, 227

Irreversibility: 97–99, 161

Printed in the United States
by Baker & Taylor Publisher Services